CONTENTS

Introduction		2
1	Political, Cultural and Military Conditions in Egypt and Iraq	3
2	Political, Cultural and Military Situation in other Arab Territories	7
3	Aviation in Arabia in the Immediate Pre-war Years	12
4	The Royal Iraqi Air Force in the Late 1930s	18
5	Egypt, from EAAF to REAF	32
Bibliography		70
About the Author		72

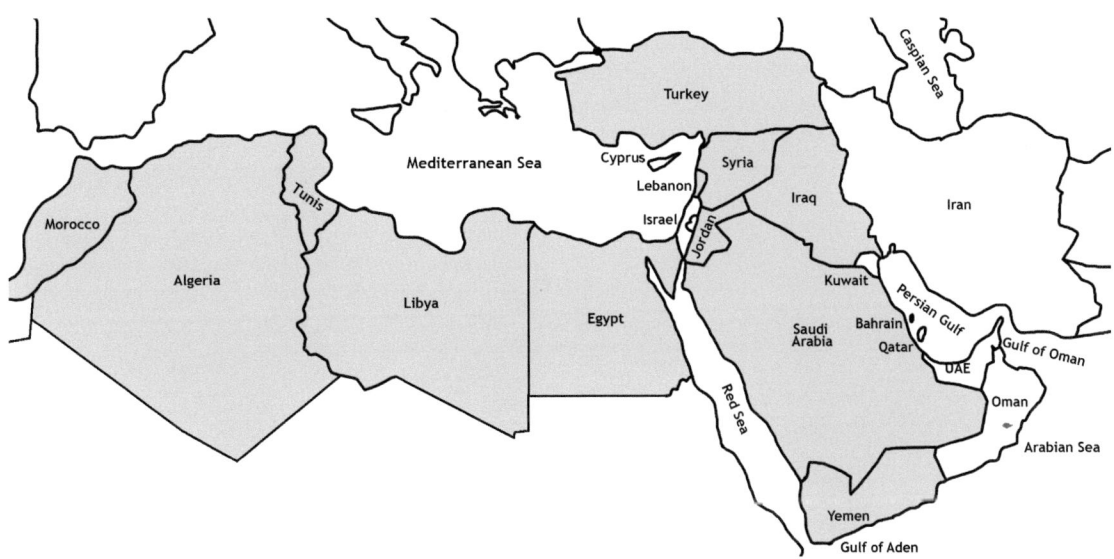

Helion & Company Limited
Unit 8 Amherst Business Centre, Budbrooke Road, Warwick CV34 5WE, England
Tel. 01926 499 619
Email: info@helion.co.uk   Website: www.helion.co.uk   Twitter: @helionbooks   Visit our blog http://blog.helion.co.uk/

Published by Helion & Company 2021
Designed and typeset by Farr out Publications, Wokingham, Berkshire
Cover designed by Paul Hewitt, Battlefield Design (www.battlefield-design.co.uk)

Text © Dr. David Nicolle, Air Vice-Marshal Dr.Gabr Ali Gabr (EAF ret.) and Tom Cooper, with Waleed Miqaati and Nour Bardai 2021
Photographs © as individually credited
Colour profiles © Tom Cooper, Peter Penev and Goran Sudar 2021
Maps © Tom Cooper 2021

Every reasonable effort has been made to trace copyright holders and to obtain their permission for the use of
copyright material. The author and publisher apologise for any errors or omissions in this work, and would be grateful
if notified of any corrections that should be incorporated in future reprints or editions of this book.

ISBN 978-1-914377-23-5

British Library Cataloguing-in-Publication Data.
A catalogue record for this book is available from the British Library.

All rights reserved. No part of this publication may be reproduced, stored in a retrieval system, or transmitted, in any form, or by any means,
electronic, mechanical, photocopying, recording or otherwise, without the express written consent of Helion & Company Limited.

For details of other military history titles published by Helion & Company Limited contact the above address, or visit our
website: http://www.helion.co.uk. We always welcome receiving book proposals from prospective authors.

DEDICATION

For Air Commodore Abd al-Moneim Miqaati (8 July 1904 to 5 April 1982) in memory of a meeting in Groppi's in 1970.

"By this, he seemed to mean, not only that the most reliable and useful courage was that which arises from the fair estimation of the encountered peril, but that an utterly fearless man is a far more dangerous comrade than a coward." (Herman Melville [abridged by A.E.W. Blake], Moby Dick, or the White Whale, London 1926)

INTRODUCTION

THE ARABS AS BYSTANDERS

If the Imperial Airways Handley Page HP 42 had been the epitome of luxurious air travel in the early 1930s (see Volume 3), the Short Empire flying boats took over that mantle in 1937. Initially they flew a regular service from Southampton in England to Kisamu in Kenya, with various stops; passengers then proceeded to South Africa by land-based aeroplanes. A later variation of the route saw flying boats landing on the Nile in the middle of Cairo. As before, passengers spent a night in the fashionable Shepheard's Hotel, after which a launch would ferry them to the waiting

Imperial Airways Handley Page HP 42 (registration G-AAUC) named *Horsa*, had to make an emergency night-time landing in the Arabian desert in August 1936. Next day the stranded machine was spotted by the RAF who returned with water and provisions. The passenger and crew were then rescued. (Fleetway archive)

Handley Page HP 42 (registration G-AAUD) named *Hanno* was part of the Imperial Airways fleet and is seen here at Gwadar, an enclave on Omani territory on the coast of Baluchistan which would be handed over to the new state of Pakistan in 1958. (Author's collection)

One the flight of Vickers Type 292 Wellesleys which carried out a remarkable non-stop flight from Egypt to Australia in November 1938, establishing a new distance record of 11,526 km. (Woodroffe family archive via the author)

Amiot 143 bombers of the French Armée de l'Air's GB I-34 and GB II-34, normally based at Chartres in France landed in the deep Sahara Desert on their way to Awlif [Aoulef] in south-central Algeria, while on manoeuvres in North Africa during December 1937. (Jarrige collection)

Empire flying boat. On-board they would enjoy good meals and often better wine as the majestic aeroplane droned south, with intermediary stops on rivers or lakes, to Durban.

This seemingly idyllic situation would be overturned by the outbreak of the Second World War in September 1939. Even before this date, however, Germany's Nazi government made considerable efforts through diplomacy and propaganda to convince Arab peoples and their governments – where such governments existed beyond French and British control – that Germany had no territorial ambitions in their part of the world.

1
POLITICAL, CULTURAL AND MILITARY CONDITIONS IN EGYPT AND IRAQ

EGYPT

King Fu'ad of Egypt died on 28 April 1936 and was succeeded by his popular sixteen-year-old son Faruq who came to the throne in an atmosphere of hope. A major role in the youngster's education had been played by the strongly nationalist General Aziz al-Masri who continued to have a strong influence upon Egypt's new king. Widely seen as religious and straightforward, Faruq had the advantage of being regarded as a real Egyptian, unlike his Turkish forebears; a view strengthened by his marriage on 20 January 1938, to an Egyptian rather than Turkish, Circassian or Albanian woman: Safinaz Zulficar, daughter of a prominent Egyptian lawyer Yussuf Zulficar. Adopting the name of Farida, she would give King Faruq three daughters.

The young monarch was fortunate in the timing of his accession which was followed four months later by the signing of a new treaty of alliance between Egypt and Great Britain on 26 August 1936. This, it was widely believed, would put the relationship on a more equal footing. Sadly, such hopes would be followed by bitter disappointment and decades of mistrust. King Faruq the Pious, as he liked to be called, soon demonstrated his patriotic aims by insisting upon a significantly more active foreign policy than that of his father. However, Faruq would prove to be a flawed character and tragic figure.

It was during the final years of Fu'ad's reign and the early years of Faruq's that the Egyptian Air Force was created in an atmosphere of optimism concerning Anglo-Egyptian relations. In fact, the new Anglo-Egyptian Treaty was largely a result of the Italian invasion of Ethiopia. Egypt clearly could resist a comparable fascist assault and while a direct threat to Egypt seemed unlikely, the Italians were believed to covet the Anglo-Egyptian Sudan whose seizure would link their colonies in Libya to their new empire in East Africa. Writing around 1951, Mustafa Nahas, a leading figure in the Wafd Party who served as Egyptian Prime Minister five times between the late 1920s and early 1950s, maintained that this caused real concern within the Egyptian political elite. Meanwhile, there was growing rivalry between Egypt and the new Arab monarchy which Britain had established in Iraq.

A poor quality but rare photograph of the Meindl/Van Nes A-VII monoplane, also known as the M7, constructed in Ethiopia for the Ethiopian Air Force by Ludwig Weber. It had an air-cooled Walter NZ 80 engine and high-lift flaps which also served as ailerons. After the Italians conquered Ethiopia, this unique machine was preserved in the Italian Air Force Museum at Vigna di Valle. (Author's collection)

Following the new Anglo-Egyptian Treaty, strenuous efforts were made to re-equip and retrain Egypt's armed forces. Intakes of officer cadets not only increased in size but also included significant numbers of middle-class youngsters whose specifically Egyptian, rather than Turkish, Circassian or Albanian, family backgrounds would previously have limited their chance of selection. Amongst them were Gamal Abd al-Nasser and several colleagues in what would later become the Free Officers Movement.

Recruitment into the non-commissioned ranks had not been a problem because there were plenty of volunteers in a poor country where military service provided regular pay and medical services. On the

While flying home to Kenya after taking part in the King's Cup Air Race in 1937, Brigadier General Lewin made a forced landing in his Miles Whitney Straight in the vast southern swamps or *Sudd* of what was then the Anglo-Egyptian Sudan. The aeroplane inevitably overturned and the Brigadier and his wife were lucky to be found by an RAF aircraft. They then had to wait until a patrol of the Sudanese Egyptian Army rescued them. (Author's collection)

Muhammad Sidqi Mahmud al-Milaigy [3rd from viewer's right in the centre file] with another unnamed Egyptian [3rd from the left], when both were student pilots in E Flight, No. 32 Course, at the RAF's No. 4 FTS Abu Suwayr from July 1935 to April 1936. (Squadron Leader C. Wright photograph)

An Armstrong Whitworth Atlas of RAF No. 4 FTS Abu Suwayr photographed in flight by Hassan Tawfiq, another REAF student at the time. (Mona Tewfik collection)

Local Egyptian labourers helping recover a crashed RAF Hawker Hart Trainer (number K4912) in 1937. (Albert Grandolini collection)

other hand, the conscription system required for an expanded army proved unfair because families with money could purchase exemptions. Another problem, which remained after expansion, was the differing attitude that the British had towards Sudanese troops, whom they regarded as unsophisticated but loyal, and Egyptian troops who were clearly more independent-minded. Furthermore, the Egyptians' loyalty was to their own king rather than to white officers from a foreign imperial power.

By 1939 the Egyptian Army, the recently created Air Force and the Egyptian Coast Guard (there was currently no Egyptian Navy) were still structured, uniformed and equipped along British lines. However, there remained serious deficiencies in training and weaponry because many in the British government and the British Embassy still doubted the wisdom of increasing Egyptian military strength. Nevertheless a "Joint Plan for the Defence of Egypt" was agreed by the British commander in Egypt, Henry Maitland Wilson, and the Egyptian Minister of Defence shortly before the outbreak of the Second World War. Under this arrangement, Egyptians would patrol the Libyan frontier, provide garrisons at Marsa Matruh and Alexandria, guard railways and establish a mobile force equipped for desert warfare to defend the south-western approaches of Cairo. They would also man coastal defences positions, provide anti-aircraft artillery for strategic bases and carry out anti-sabotage duties. The plan would stretch the small Egyptian Army to its limit but would free British troops for active operations in Egypt and elsewhere.

An Egyptian Avro 626 of what had been the Egyptian Army Air Force, now the Royal Egyptian Air Force, around 1936. It was probably on the beach at Marsa Matruh, which served for many years as the REAF's closest airfield to the Libyan frontier. (Albert Grandolini collection)

Hawker Persian Audax number 301, on the ground before delivery to Iran but shown with bombs attached to its under-wing bomb racks. (Albert Grandolini collection)

Hawker Persian Fury number 208 of Imperial Iranian Air Force, probably photographed over England before being delivered to the Middle East. This example appears to be one of the six aircraft ordered in 1934 which had Bristol Mercury VISP radial engines with two-bladed rather than the three-bladed propellers of the earlier example. A few Persian Furies would be used by the Iranians as late as 1949. (Albert Grandolini collection)

IRAQ

During the later 1930s, Iraqi politicians were increasingly split between those regarded as pro-British, including Nuri al-Sa'id, and those considered anti-British such as Rashid Ali al-Gaylani (Kaylani). Meanwhile, sentiment in the armed forces, including the new Royal Iraqi Air Force, was heavily weighted against what was perceived as a continuing and barely disguised British occupation. The passionate nature of these divisions and the fragility of the somewhat artificial Hashemite Monarchy of Iraq would result in no less than five military coups between 1936 and 1941. However, Iraq's problems were not based solely upon its relationship with the British Empire, as there was an increasing tendency for the higher military command to resort to coups to pressure Iraqi governments into conceding to the army's demands. It was a baleful habit that would cause the country difficulties throughout the rest of the 20th century.

Muhammad Abd al-Hamid Abu Zaid [2nd row far right] with fellow cadets at the Egyptian Army's Military Academy in Cairo in 1938. He would then join the REAF and become the best-known Egyptian pilot during the Palestine War of 1948. (Abu Zaid family collection)

2
POLITICAL, CULTURAL AND MILITARY SITUATION IN OTHER ARAB TERRITORIES

The rise to power of Adolf Hitler's Nazi Party in Germany in 1933, along with increasing anti-Semitism in many countries across Europe, North America and beyond, led to an increase in a Zionist migration to Palestine. Here relations between the indigenous Arab population and Jewish settlers steadily worsened. This period also saw a number of contacts between the viciously anti-Semitic Nazi regime and some marginal or extremist groups within the Zionist movement. These are still so controversial that most historians – and politicians – steer well clear or deny that they happened. One such extraordinary set of meetings, intended to facilitate the emigration of German Jews to Palestine, resulted in the so-called Heskem Haavara or "transfer agreement" between the Zionist Federation of Germany, the Anglo-Palestine Bank under the direction of the Jewish Agency for Israel, and elements of the German government. Surprisingly perhaps, this peculiar arrangement reportedly enabled approximately 60,000 German Jews to reach Palestine between 1933 and the outbreak of the Second World War in 1939.

Other meetings would follow which, in the light of the subsequent Holocaust or "Shoah", now appear totally bizarre. In 1937, for example, a mission led by Herbert Hagen included none other than Adolf Eichmann who would later be executed in Israel on 1 June 1965. It was sent to meet Zionist leaders in the Middle East to discuss how an "accommodation" could be reached that would solve what anti-Semites called "the Jewish question". Landing at Haifa, the Nazi delegation obtained a transit visa from the British authorities before travelling to Cairo where they met a senior member of the Haganah, or Zionist paramilitary force in Palestine. Here Hagen and Eichmann reportedly discussed methods of boosting the emigration of German Jews to Palestine, this being seen as beneficial by some in the Zionist leadership as well as many in the Nazi leadership. Nothing seems to have come of the meeting, yet there were a few extreme Zionists who continued to regard the Nazis as what one historian has called "potentially useful partners", even after the start of the Second World War but before the Nazis launched their horrific "Final Solution" in January 1942. In the spring of 1938 Hitler himself spoke publicly in support of the emigration – or more truthfully the expulsion – of German Jews to the Middle East and even, it is claimed, the formation of a new state to be their home in the area. In general, however, Nazi ideology opposed such a Jewish state anywhere.

Meanwhile the Palestinian Arabs had been growing increasingly concerned about their own future. This tinderbox was ignited in November 1935 when British colonial police killed Shaykh Izz al-Din al-Qassam near Jinin. This was followed by a major Arab uprising against both British rule and Zionist settlement, lasting from 1936 to 1939. Starting with a strike and some sporadic armed attacks, the Great Rising, as it came to be known, gradually grew more coordinated and organised. Strategic British installations like railways and the Trans-Arabian Pipeline (TAP) were primary targets. There were also smaller attacks on Zionist settlements and Jewish neighbourhoods.

In 1937 LOT, the Polish national airlines, inaugurated a regular service between Warsaw and Lydda using Douglas DC-2 aircraft. It was regarded as one of the longest routes "in Europe" and linked the Zionist settler community in Palestine with the substantial Jewish community in Poland. (Author's collection)

Palestinian irregulars in 1938, with a flag bearing both a cross and a crescent to emphasise the unity of the country's diverse indigenous population. (Albert Grandolini collection)

For a while the trouble died down while the British government's Royal Palestine Commission, commonly known as the Peel Commission, studied the situation, eventually recommending that the country be partitioned between Arabs and Jews. This suggestion was rejected by the horrified Palestinian Arab leadership, whereupon the Great Revolt flared up again. This time the British response was ruthless to the point of brutality, increasing the number of British military personnel, and imposing "administrative detention" (a form of detention without trial or formal charge which is still employed by Israeli occupation authorities in what remains of Palestine). It was used against any person suspected of dissent. Curfews and the demolition of the

A British Army armoured vehicle, apparently made of a small commercial vehicle with additional armoured plates and with flanged railway wheels, in front of a train of the Palestine Railways system in 1938 (Albert Grandolini collection)

fighters at the high point of the insurgency in autumn 1938. Such cooperation was not officially admitted by the British, who did not officially recognise the Haganah, and was largely carried out under the umbrella of the recognised Jewish Settlement Police. At the same time an extremist group broke away from the Haganah – though not becoming entirely separate – to form a thoroughgoing terrorist organisation called the "Irgun" or "Etzel". Its policy was one of savage retaliation for any perceived attack, meeting violence with greater violence, resulting in attacks against markets, cafes and other places where Arabs gathered.

Although the Great Revolt failed, it is widely credited with marking the birth of a clear Arab Palestinian identity. It also resulted in the British government setting up another commission, this time led by Sir John Woodhead, a highly experienced former civil administrator in British India. In 1938 the Woodhead Commission issued its own

houses of "rebels" and their families were also introduced by the British and similarly remain in use.

The Haganah Zionist paramilitary organisation supported British efforts against what were estimated to be 10,000 Arab

A motorised patrol of the British Army in Palestine during 1938. Being occasionally shot at by both Jewish Zionist settlers and the indigenous Palestinian Arab population, as well as being subject to criticism from around the world, this increasingly became one of the most unpopular postings in the British Army. (Albert Grandolini collection) report, rejecting the previous Peel Commission's proposal for a simple partition. This was done on the grounds that such a partition would be impossible without a massive, forced removal of Arab populations from areas allocated to the Zionists – an idea which the British had already rejected as inhumane. Instead, the Woodhead Commission put forward three more complex plans, all of which involved a greater or lesser continuation of the British Mandate in various parts of the country, but these were rejected as impractical by a frustrated British government which now had its attention focussed upon the growing threat of war with Germany.

The parasol wing Nieuport-Delage NiD 62 fighter was obsolescent by January 1939, when this machine of the French Armée de l'Air was photographed at al-Blidah [Blida] in Algeria. It had been introduced into service back in 1931 but was now being used as an advanced fighter trainer. (Andre Molto via Jarrige)

Indeed, the British government virtually threw up its hands in despair, issuing a new policy statement which stated that: 'the political, administrative and financial difficulties involved in the proposal to create independent Arab and Jewish States inside Palestine are so great that this solution of the problem is impracticable'. It was followed by the St James Conference in London from 7 February to 17 March 1939, where both the Zionists and the Palestinian Arabs supported by neighbouring Arab states, all rejected the existing proposals.

The Farman F.221 was an early version of the F.220 series which served as one of the French Armée de l'Air's heavy bombers during the mid- and later 1930s. Later versions had retractable undercarriages. This old design nevertheless soldiered on after the fall of France in both Vichy French and Free French Air Force colours. An example is seen here over Algeria before the Second World War. (Albert Grandolini collection)

Less than six months later the world was plunged into the Second World War.

Elsewhere, following riots in Tunisia in 1938, the French authorities tried to suppress the nationalist Neo-Destour Party which instead sought and found support from Fascist Italy. Meanwhile Habib Bourguiba, one of the leaders of the Neo-Destour and subsequently the first president of independent Tunisia, was imprisoned in Marseilles where he remained during the Second World War.

Back in Palestine, the Arab Revolt of 1936–1939 had put huge strain on the Palestine Police Force which, though mixed in its recruitment, remained a largely Arab force. As a result, the British accepted the need to establish other forces, theoretically to work in conjunction with the Palestine Police but in practice often acting autonomously when not actually in opposition to it. These included the Jewish Supernumerary Police, the Jewish Settlement Police and the ominous sounding British-Jewish Special Night Squads.

During the British mandate, police posts had been established throughout Palestine, but from the late 1930s onwards a new network of standardised fortified posts was constructed. Designed by the British engineer Charles Tegart in 1938, they incorporated lessons learned during a similar anti-British insurgency in India where Tegart had been Police Commissioner in Calcutta (Kolkata). Known as Tegart Forts – and also by less repeatable names – many can still be seen across Israel and the Occupied Palestinian Territories. Similarly, "Tegart's Wall" was a barbed wire fence reinforced by police forts and pillboxes close to the Lebanese and Syrian frontiers. Unfortunately, its similarities with "Graziani's Fence", the massive, barbed wire obstacle built by Fascist Italy along the frontier between Libya and Egypt, did not go unnoticed.

By the time Frederick G. Peake, or Peake Pasha as he was affectionately known, retired in 1939, he had risen to the rank of Major General in what was by then the Army of the Amirate of Transjordan, better known as the Arab Legion. He was succeeded by John Bagot Glubb, or Glubb Pasha. Records show that Peake's

Major General Frederick G. Peake (Peake Pasha) retired as commander of the Transjordanian Arab Legion in 1939. His De Havilland DH 60M Moth (registration G-ABMX) was sold to C. Katz of the Palestine Flying Services and its fuselage was still being used for ground instruction by the Israelis in 1948, as seen here. (Author's collection)

De Havilland DH 60M Gipsy Moth (registration G-ABMX, c/n 1686) was sold to C. Katz of Palestine Flying Services based at Lydda in Palestine, seemingly before Peake's retirement from the Transjordanian Army. The Moth was, or perhaps already had been, badly damaged when it swung off the runway at Lydda airport on 25 June 1938. Whether it was repaired to flying condition is unclear, as the aeroplane's Certificate of Airworthiness expired on 12 June 1939 and its registration was cancelled 12 days later.

The Moth, still with its British civil registration, was then used as an instructional airframe at Tel Aviv aerodrome (presumably Sde Dov) and still existed in 1948 when this airfield came under attack from the Royal Egyptian Air Force (REAF) during the Palestine War. A surviving photograph from around this time shows that it still had the chequered tailfin, probably red and white, it had when flown by Peake Pasha. This design was subsequently used as a squadron identification marking in the Royal Jordanian Air Force.

Among other, smaller Arab military units established or restructured by the British during the inter-war years was the Royal Army of Oman. Officially its origins can be traced back to the establishment of the Muscat Garrison in 1907, but this was considerably enlarged in 1921 when it was renamed the Muscat Infantry. Similarly, the Kuwait Army, officially established in 1949, can trace its origins back to 1919 when Britain encouraged the Shaykh of Kuwait to raise small numbers of cavalry and infantry for desert patrol work. However, the Trucial Oman Levies, the predecessor of the Army of the United Arab Emirates (established in 1971), was only created in 1951. Prior to that the seven separate emirates of what the British called the Trucial Coast each had their own small tribal forces.

Meanwhile, following Fascist Italy's ruthless crushing of Libyan resistance (see Volume 3), Mussolini's government set about trying to win Libyan "hearts and minds". This succeeded to a limited though surprising degree, given recent events, and was symbolised by the Italian dictator's "Sword of Islam" campaign. To start with, Italian military forces in Libya were reorganised under a new High Command for North Africa. Benito Mussolini and his retinue then set sail from Gaeta aboard the cruiser *Pola*, arriving in Tobruk on 12 March 1937 for a full-scale and widely publicised visit.

Initially focussing upon Italian settler villages, and watching a performance of the Greek tragedy *Oedipus Rex* by Sophocles in the recently cleared Roman theatre at Sabratha, Mussolini than travelled to the Oasis of Bugara outside the city of Tripoli. Here, on 20 March 1937 in front of reported 2,600 "Arab horsemen", he received a gilded sword from a Berber leader named Yusuf Kirbish, a supporter of the Italian occupation. At this lavish but widely lampooned ceremony the fascist dictator proclaimed that Italy

Paratroopers of the Italian Army's 1st Libyan Infantry *Divisione Sibille*. These locally recruited colonial troops formed the Arab world's first unit of paratroopers and although they fought during the forthcoming North African Campaign, they were not called upon to drop into enemy-held territory. (Albert Grandolini collection)

was the friend and protector of Islam, with the implication that fascism would be the Arabs' ally against British and French oppression.

Mussolini's words were followed by further restructuring of Italian forces in Libya during which Italy demonstrated its faith in the previously shaky loyalty of Libyan colonial troops by establishing the Arab world's first unit of paratroopers. It would form part of the 1st Libyan Division, also known as the Divisione Sibille after its commander, General Luigi Sibille. Recruited from Arabs and Berbers, this battalion of Ascari del Cielo was the first unit of paratroopers in the Italian Army, the term Ascari reflecting the fact its men were indigenous colonial soldiers. The Ascari Battalion formally came into existence on 22 March 1938, consisting of around 300 Libyans plus 30 Italian officers and NCOs. They were commanded by Major Goffredo Tonini and were stationed at the airport of Castel Benito near Tripoli. Italians were soon also trained as paratroopers and together they become the 1st Parachute Regiment of Fanti dell'Aria.

The three-engine, fixed undercarriage Savoia-Marchetti SM 81 selected as Italy's first paratrooper transport had been designed as a bomber and was basically a militarised version of the SM 73 airliner. Thus, the Ascari del Cielo trained on off-white SM 81s of the Regia Aeronautica's 15th Stormo, based at Castel Benito. Here, on 16 April 1938 after four weeks of training, Major Tonini's 300 Libyan volunteers made their first coordinated drop from 24 SM 81s, watched by the Governor-General of Libya, Marshal Italo Balbo. A few days later another drop was made at night.

Training in these early days of airborne warfare was highly dangerous and the first of an

This Soviet-built Tupolev SB-2 Katiuska high-speed bomber, individual code number 35, was one of those Spanish Republican Air Force machines which escaped to French-ruled Algeria at the end of the Spanish Civil War. It was photographed at Mustaghanim Tighdit (Mostaganem Tigditt) aerodrome in March 1939. The civil war formally came to an end on 1 April that year. (Pierre Blanchet via Jarrige archive)

One of the 20 Junkers Ju 52/3m transports which, with civilian and military air and support crews, Nazi Germany sent to aid General Franco's rebellion against the Spanish government in 1936. They were promptly given Spanish markings. Beneath the wings sit some of those North African troops who, recruited in Spain's Protectorate in northern Morocco, proved vital to Franco's cause. (H.J. Nowarra collection)

A Northrop 1D Delta amongst other aircraft of the Spanish Republican Air Force which escaped to French-ruled Algeria when General Franco's fascist army finally won the Spanish Civil War. They are seen here at La Senia airport outside Qahran [Oran], probably in 1939. (Jean-Marc Paous via Jarrige)

One of the Junkers Ju-52/3m airliner-transports sent to support General Franco's Nationalist Army at the start of the Spanish Civil War, these aircraft initially retained their German civil registrations, in this case D-ATRN, as well as Spanish Air Force markings and numbers on their fuselages, plus nationalist crosses on their rudders. (Heinz Nowarra archive)

estimated 20 casualties suffered by Libyan paratroopers before the outbreak of the Second World War was Ugashi Muhammad Ibn Ali. Along with other Arab and Berber ascaris of the Italian Army, the uniforms of the Libyan paratroopers adopted the five-pointed star of the metropolitan Italian Army when the four provinces of their homeland were incorporated into the Kingdom of Italy in January 1939 and the status of Libya became comparable to that of French-ruled Algeria. In fact, the Italian Army was so proud of its Libyan paratroopers that they were presented to Reichsmarschall Hermann Goering when he visited Libya on 9 April 1939.

Further west, in the Spanish Protectorate in northern Morocco, the Spanish Army had long enlisted local soldiers, large numbers of whom came from the same Berber tribes that had for decades caused major problems for the Spaniards (see Volumes 1 to 4). When the Spanish Civil War erupted in July 1936, the leaders of the Spanish Nationalists' attempted coup found that their existing forces in southern Spain were in danger of defeat. They needed to rapidly transfer pro-Nationalist troops from northern Morocco to the Spanish mainland but the seas between were dominated by pro-government, or Spanish Republican, warships. The answer lay in an airlift, but the Nationalists lacked suitable aeroplanes, so Nazi Germany promptly offered the services of 20 Junkers Ju 52/3m transports, along with air and support crews drawn from Lufthansa and the Luftwaffe. Despite being fired upon by Spanish government warships, notably the battle-cruiser *Jaime I* which had highly effective anti-aircraft guns, the airlift was a significant success. Approximately 15,000 Moroccan and Spanish troops were flown across the Straits in what could be seen as a reworking of the Arab-Berber-Islamic invasion of Spain over 1,200 years earlier. Then, a small fleet of Christian Byzantine ships carried the seasick invaders from Morocco to southern Spain; this time a fleet of Christian Nazi aircraft carried the airsick North African soldiers over much the same waters.

3
AVIATION IN ARABIA IN THE IMMEDIATE PRE-WAR YEARS

Normally the aeroplanes of the recently formed Saudi Arabian Air Force were overhauled during the month of Ramadan, when Saudi pilots and ground crew were unable to work during the daylight hours because they were fasting. A lack of sufficient spares and lubricants nevertheless proved a problem, while the resident Italian technical support mission were also having difficult drinking local water. Instead, they had been urged to drink bottled water, which was expensive. Their commander, Captain Giovanni Battista Ciccu then became the target of a campaign of rumours directed against all Italians. It is believed to have been started by Italy's rivals, notably the British. As a result, Ciccu had to be extremely careful of what he said and did, especially as some Arab pilots of the Saudi Air Force were not actually Saudi citizens. Any "wild" behaviour by these men reflected badly, if unfairly, on the Italians.

According to a report in October 1936, the design of the Saudi Air Force pilots' uniforms had now been decided upon. It was essentially the khaki of an Italian colonial pilot officer, plus Saudi headgear. Captain Ciccu also suggested that, to further distinguish Saudi from Italian uniform, the five-pointed stars used by other Saudi armed forces to indicate rank, be used along with a green sash instead of Italy's blue. A golden wing was applied to the epaulette, replacing the sceptre used by Italian pilot officers.

Caproni Ca.101bis (registration I-ABCC) before it was sold to the Saudi Arabian government in August 1937. This aeroplane had previously been operated by Nord Africa Aviazione and then Ala Littoria, both based in Benghazi, in Libya. (Lennart Andersson archive)

Lieutenant Col. Carlo Tempesti, the first leader of the Italian Air Force Mission to Saudi Arabia in 1936. He soon had to return to Italy due to a serious illness and was replaced by Capt. Giovanni Battista Ciccu. (Author's collection)

Lieutenant Col Renato Ciancio was sent to Saudi Arabia in April 1937 as the new head of the Italian Air Force Mission. He replaced Giovanni Battista Ciccu who had fallen sick. (Author's collection)

Meanwhile, service in the expanding Saudi Air Force was becoming amongst members of the Saudi elite and on 5 November another group of six youngsters started their training to become pilots.

The three-engine Caproni Ca.101bis had already proved effective as a bomber over Ethiopia and it would provide the Saudi Air Force with its first real offensive capability. However, the type's arrival was delayed because of the Arabian peninsula's brief rainy season and a lack of suitable hangars, even though Italy offered more than one type of metal or mixed construction hangar. The Caproni Ca.101bis was further delayed because they needed complete overhaul in Italy, having come from the former Nord Africa Aviazione airline based in Libya, which had itself been absorbed by Ala Littoria on 1 August 1935. Though the Ca.101bis had proved strong and reliable in the harsh conditions of Italy's North African colony, the engines similarly required total overhaul to zero-hours condition.

The first machine to be ready in mid-September had the Italian civilian registration I-ABCC and may also have still carried the Italian military identification number MM20553, having previously served in the Italian Air Force. Despite now being ready, I-ABCC could not set out because of delays obtaining overflight permission from the French, British and Egyptian governments, in turn causing problems for Captain Ciccu in Saudi Arabia. Eventually the Ca.101bis left Rome on 23 October, then flew via Palermo, Tunis, Tripoli, Benghazi, Sollum, Marsa Matruh, Cairo-Almaza, Luxor, Wadi Halfa, Karima, Khartoum, Kassala and Massawa to Jiddah where it landed at 9:40 a.m. on 4 November. Ciccu made the first dual control training flight the following day with six Saudi pilots, though the formal handover of the machine was not until 8 November. This second-hand aeroplane, modified as a passenger transport, now made five publicity flights carrying various Italian and Saudi dignitaries. Thereafter flights were made every day from the new Kandara aerodrome outside Jiddah, which had been home to the Arabian School of Aviation since late September.

The assembling of the single-engined Ca.100s and the delivery of the first Ca.101bis now being complete, the original two Italian mechanics in Saudi Arabia left for home on 11 November, along with the tri-motor's delivery crew. Meanwhile Ciccu and the Russian pilot Nikolai Naidenov set about instructing the would-be Saudi pilots on two of the Caproni primary trainers and the four-seater Caudron C510 Pelican given to the Saudi Air Force back in 1935.

The next year, 1937, would see important changes, the first of which was when Ciccu was replaced by Lieutenant Colonel Renato Ciancio as head of the Italian Air Force Mission to Saudi Arabia in January. Ciancio was younger but senior to Ciccu and seems to have reached Saudi Arabia the following April. It was probably around January 1937 that Italy offered to send engineers to continue the construction of Jiddah aerodrome, now that Saudi Arabia's contract with Britain had expired. Apart from being an

Prince Faisal Ibn Abd al-Aziz, the third son of King Abd al-Aziz of Saudi Arabia, with his entourage and Capt. Giovanni Baptista Ciccu of the Italian Military Mission. Italy's effort to sell aircraft and training to the Saudis was at this stage very successful. (Ciccu archive)

Capt. Giovanni Battista Ciccu, who arrived in Saudi Arabia in June 1936, soon became aware of the difficulties of establishing an effective Air Force in the desert kingdom. He is seen here leaning into a high wind in front of a Saporiti metal clad hangar. Traditional canvas hangars had not proved to be very suitable in such conditions. (Ciccu archive)

important military and training centre, the Saudi air base at Jiddah was often used for publicity, as when two Saudi pilots escorted Prince Faisal's cavalcade of cars into the port city on 4 January 1937 while two others flew over the royal palace at low altitude. Aeroplanes would similarly be used to impress Muslim Hajjis or pilgrims as they arrived in Jiddah from across the world, dropping propaganda leaflets as the pilgrims set off on the final stages of their Hajj to the holy sites in Mecca and Madina in the second half of February 1937.

Around the same time, a metal hangar had been obtained from the Italians. Measuring 52 by 33 by 7 metres, it was successfully erected by 23 February, despite some parts having been lost when a dhow sank outside Jiddah harbour as the hangar's pieces were brought ashore from the steamship *Alberto Treves*. This was just in time for the arrival of the other two Caproni Ca.101bis, with Italian civil registrations I-ABCI and I-ABCK (c/n 3349 and 3351), which reached Saudi Arabia on 27 March. Both were equipped with wireless and, like the first of this type in Saudi Arabia, carried civil registration numbers in an attempt to make it easier to obtain overflight permits. Although they had been so delayed, once the machines were in Saudi Arabia the training of aircrew became much easier, despite the fact that the Italian instructors considered navigation more problematical than it had been in Libya.

When Lieutenant Colonel Ciancio arrived, he appears to have brought another technician with him so that henceforth there were four rather than the previous three, including an engineer, a rigger and a radio operator, who were rotated at regular intervals. Efforts were also made to keep the Westland Wapitis operational, using spares obtained from the British. These aeroplanes remained the responsibility of Naidenov, Maximov and a second recently arrived Russian mechanic. The Russians were probably also responsible for taking a Siddeley Puma engine from one of the old De Havilland DH 9s and using it to power a motorboat.

A published photograph from this period shows King Abd al-Aziz Ibn Sa'ud getting out of one of the Caproni Ca.101bis, but it is unclear whether he actually flew in it. In late April 1937 the Crown Prince and one of the King's brothers were certainly taken up by several Saudi pilots, along with other members of the Saudi royal family. Then in April or May, the Saudi Arabian Mining Syndicate donated their Bellanca CH-400 Skyrocket to the Saudi government, having failed to sell it to them.

Not surprisingly, the Italian Mission urged the Saudi government to purchase more Italian aeroplanes. To give the

Air Force a proper military capability it was suggested that single-engine Ca.111 or three-engine Ca.133 bombers be bought, along with another three to five Caproni primary trainers. Funds for these were available through the efforts of the Arab Aeronautical Society which had raised over 13,000 lbs in gold bullion by May 1937. This was, of course, before Saudi Arabia became wealthy as a major oil exporter. Nevertheless, nothing came of these efforts though it has been suggested that an Iraqi Air Force officer, the same Musa Ali who had experience with Italian SM 79B and Breda Ba 65 aircraft, became involved in the discussions in 1938. In October 1937 Captain Giovanni Battista Ciccu returned again to take over as head of the Italian Mission, and before the end of the year he and the Saudi government decided that in future all flying students in the Jiddah flying school must be Saudi subjects. For the Saudi Air Force, the final event of 1937 was a flying display by two Ca.101s, the Caudron and the Bellanca, on 13 December.

During 1938 normal duties were continued, patrols being flown despite the unavailability of correct octane fuel. Meanwhile efforts were made to improve the drainage of Jiddah aerodrome which occasionally flooded at high tide, and to build a proper runway. This was despite the difficulty of getting materials from Italy, Eritrea or Italian Somalia. Improvements to other airfields at Yanbu and al-Wijh were similarly considered while, in an effort to avoid the Saudi Air Force becoming entirely dominated by Italy, the French government donated a Caudron Simoun high-speed low-wing monoplane (French civil registration F-AQLY, c/n 139/7758/90).

By May 1938, Sa'id al-Kurdi had been appointed as the new Officer Commanding Saudi Aviation. Also during May, Captain Ciccu, Nikolai Naidenov and two young Saudi pilots flew the three Caproni Ca.101b and the Bellanca to Yanbu in a formation exercise. The Saudi Air Force now had, in addition to these machines, four Wapitis – two of which were unserviceable – just two Caproni Ca.100 trainers, the new Caudron Simoun and the old Caudron Pelican. The Russian pilot Naidenov had already trained Abdullah Mandili and Tarabzanli Sadaka to fly the latter.

One of the Caproni Ca. 101 bis transports (Italian registrations I-ABCC and I-ABCI) sold to the Saudi Arabian Air Force in August 1937, and photographed in Saudi Arabia while still having the Italian fascist symbol painted on the sides of its nose. Amongst the Saudi personnel are several Europeans who are believed to include the crew who flew the Ca. 101 to Arabia in March 1937. (Lennart Andersson archive)

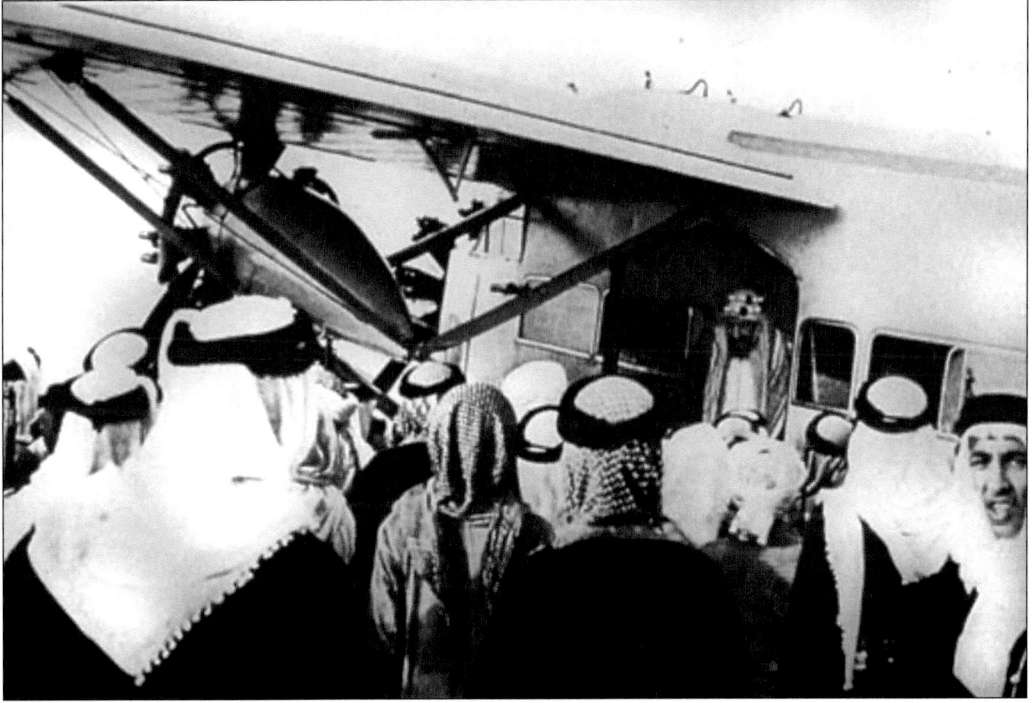

Saudi King Abd al-Aziz of Saudi Arabia emerging from one of the Caproni Ca 101bis aircraft delivered to the Saudi government in March 1937. At this point the aeroplane still had its Italian civil registration (I-ABCI) and the *Fasces* insignia. (Lennart Andersson archive)

Another Wapiti had been cannibalised for spares and was not expected to fly again.

In a change of policy, the Saudi government now decided to send Abdullah Mandili, who was considered the country's most talented pilot, to Egypt to be trained as a flying instructor rather than sending him to Italy. This was probably for political reasons. Mandili would subsequently be joined by Salih Alim, though the surviving documentation seems to write his name as Salih Atib. Saudi Arabia was almost certainly trying to strengthen its military relations with other Arab countries and in 1938 rumours were circulating about the country inviting an experienced Iraqi Air Force officer to assist in a reorganisation of Saudi armed forces. The officer in question was again Major Musa Ali, the commander of Iraq's Mu'askar al-Rashid aerodrome. To the Italians, however, this looked like a ploy whereby the British could – through Iraq – undermine Italy's dominant position in equipping and training

Caproni Ca.101bis (registration I-ABCK) around the time it was sold to the Saudi Arabian government, again in August 1937. The Caproni Ca.101bis used by the Saudis seem to have retained their Italian civil markings and registrations for a while, and in the case of I-ABCK this apparently included the Fasces symbols on the sides of the nose. (Lennart Andersson archive)

Westland Wapiti IIA number K1269 on its back after a taxying accident at the RAF' No. FTS Flying Training School. The Wapiti was notably difficult to taxi and would easily tip onto its nose, or indeed right onto its back, and this is just one of many such example at Abu Suwayr in the mid-1930s. This machine would later be written off in an accident at Drigh Road aerodrome outside Karachi, now Jinnah International Airport in Pakistan, on 30 October 1941. (Mona Tewfik collection)

the Saudi Air Force. In fact, Musa Ali had himself trained in Italy and had considerable experience of Italian aeroplanes, notably the SM 79B and Breda Ba 65.

In August Major Luigi Gori Savellini was sent to Saudi Arabia to assist Ciccu as commander of the flying school. Savellini was an experienced desert pilot, having served in Libya, and had recently been in charge of the Italian Air Forces' basic flying school at Siena. Indeed, Ciccu may already have been ill as he returned to Italy for health reasons on 4 November 1938. Luigi Savellini then took over as leader of the Italian Mission but did not get on well with the Saudis, perhaps having difficulty adjusting his behaviour from being a colonial master in Libya to that of a guest in Saudi Arabia. Nevertheless, Savellini did persuade the Saudis to allow him to bring all the Air Force's personnel, equipment and aeroplanes together under his authority as commander of the Italian Mission. He would then be directly answerable to the Saudi Minister of Defence. In return, Savellini would wear Saudi rather than Italian uniform. Unfortunately, it was the Italian government which now baulked at this idea, and the whole proposal stalled. Meanwhile Nikolai Naidenov overhauled the three surviving Wapitis at Taif. Relations between Naidenov and the Italians seems to have been cordial and on 7 December he taught Savellini to fly the Bellanca, followed by an old Wapiti early in 1939.

The discovery of oil in Saudi Arabia would not only change the country beyond recognition, but would change the attitudes of outside powers towards Saudi Arabia. Thus, when Savellini asked to be recalled "for consultation" early in 1939, he was told to remain where he was and to carry on with what was becoming a difficult and frustrating task. Eager to remain on good terms with the Saudi government, and aware that what had until recently been a very poor country was likely to become a wealthy one, Rome offered to sell equipment on easy payment terms early in May 1939. Italy similarly offered to train further Saudi pilots free of charge. It was to no avail and Saudi Arabia instead announced its intention to send future flying cadets to Egypt, citing the fact that training in Egypt would be done in Arabic. This would mean the almost immediate closure of the Italian-run flying school in Jiddah. The language problem was genuine, as Major Savellini

An Egyptian De Havilland DH 83 Fox Moth (registration SU-ABA or ABG) with a De Havilland DH 84 Dragon (registration SU-ABI) in the background. They were probably photographed during a flight to Saudi Arabia in late 1934 or early 1935 when Misr Airwork was trying to establish an air service between the two countries. (Ahmad Isma'il collection)

himself stated: 'for each flying exercise to be performed by the pilots, in order to make oneself understood it is necessary to resort to all manner of sounds and gestures'.

Days were now numbered for the Italian Mission in Saudi Arabia, and the last recorded flights by a member were in a Ca.100 and a Westland Wapiti on 6 March 1939. Just under a month later, on 1 April, the Italians left the country, Savellini taking a ship to Massawa in Italian East Africa, while the five ground crew sailed for Suez. Thereafter it became increasingly difficult to keep the Saudi Air Force flying and during May 1939 one of the three Russians still employed by the Saudis was told to repair the fourth Wapiti, though it remains unlikely that this machine ever flew again.

A major reason for the decline of the Saudi Air Force would, of course, be the outbreak of the Second World War, during which potential suppliers of aeroplanes and expertise had other things to worry about. Before the outbreak of war, however, great optimism accompanied the sending of Saudi pilots from the Jiddah flying school to the Misr Air Flying School at Almaza aerodrome on the outskirts of Cairo. At least seven would be studying here by June 1939. They stayed for only three months, being hurriedly recalled to Saudi Arabia on 16 September 1939, two weeks after the Second World War broke out.

Developments in Yemen during the later 1930s perhaps had marginal significance outside Yemen itself. Nevertheless, the fact that forces loyal to the Imam Yahya brought most of the country under the Imam's rule by 1934 enabled the very small number of educated urban nationalists to start asserting themselves. This was certainly not what the Imam Yahya had intended as a consequence of his policies. Most such nationalists were not even part of the Imam's own Zaydi Islamic community, being Sunni Muslims and mostly living in the coastal plain, the southern city of Ta'iz or in British-ruled Aden, rather than in the mountainous heartlands of the Imam's power. Some had studied in Cairo where they were influenced by both the Muslim Brotherhood and by exiled Algerian nationalists. Back in Yemen, those who had joined the Muslim Brotherhood now joined forces with existing urban opposition groups. They also offered their support to Prince Abdullah Ibn Ahmad al-Wazir who, though himself a Zaidi, wanted to overthrow Imam Yahya. Clearly the Imam's military successes had failed to unite the country and the deep-seated divisions within Yemen would rumble on, erupting into civil war in the 1960s, and remain fundamental to the ongoing conflict in Yemen.

Meanwhile, as part of a process of attempting to modernise Yemen's military structures, a new force called the Defensive Army was established in 1936. As a form of reserve, it theoretically consisted of a levy of all able-bodied men, including those from urban populations who had traditionally been excluded from military duties. They were supposed to receive six months initial training, to be followed by further periods spread over ten years.

The substantial and very long-established Jewish population in Yemen tended to be supportive of Imam Yahya's regime, Jewish chronicles lavishing praise upon the Imam and depicting him as the champion of justice and compassion. Certainly, Imam Yahya brought an end to the state of anarchy, lawlessness and violence which had plagued Yemen for decades, and during his long reign the Jews enjoyed relatively favourable conditions.

On the other side of the Red Sea the successful Italian invasion of Ethiopia and the historic impossibility of an alliance between Yemen and the British Empire were primary reasons why Imam Yahya signed the Treaty of Arab Brotherhood and Alliance on 29 April 1937. This was an early manifestation of what has proved to be the elusive goal of Arab Unity. It had been preceded by the "Arab Treaty" between Saudi Arabia and Iraq which, signed on 2 April 1936, stated that Yemen and other independent Arab nations could be invited to join at a future date.

Imam Yahya may even have been toying with the idea of a revived Yemeni Air Force early in 1939, when a Greek businessman from Athens, Andrea Papadopoulos, sold him three Polish-made DWL RWD-13 high-wing enclosed cabin monoplanes (c/n 259, 260 and 261), powered by 130 hp Gipsy Major engines. One of these RWD-13s was flown to Sana'a in July 1939 by a Polish pilot and was then demonstrated to the Yemeni ruler. This machine is known to have been given the registration YEMEN-2, so it is almost certain that the others were YEMEN-1 and YEMEN-3, these names probably being written in Arabic though this remains unconfirmed. They were delivered by sea, in crates, to Hudaydah but, for reasons which also remain unknown, were never accepted by the Yemeni government. In all probability this problem was caused by the outbreak of the Second World War, the German invasion of Poland preventing the three Polish aeroplanes being returned.

4

THE ROYAL IRAQI AIR FORCE IN THE LATE 1930S

Ongoing operations against what was called the "Second Euphrates Uprising" involved Nos. 2 and 3 Squadrons of the RIrqAF and resulted in further losses for them. On 15 May 1936 the No. 3 Squadron Hawker Nisr of *Mulazim Awal Tayyar* Ra'uf Shabib was shot down by groundfire, killing both the pilot and his observer, *Jundi* Tawfik Yahya, near Rumaytha, south of al-Diwaniya. The following day, and in the same area, a Nisr flown by Captain Armiya Nasir was also shot down by dissident tribesmen, killing both the pilot and his observer Abd Allah Husayn. Apparently, Nasir managed to bail out, but was too close to the ground for his parachute to fully open. No. 3 Squadron's tour of duty against this particular uprising ended on 18 June 1936, though the unit would be back in action the following year. No. 2 Squadron was operating in the Rumaytha area from 21 to 27 February and again from 10 to 18 June 1936, though without serious casualties.

Outside Baghdad, meanwhile, the newly qualified group of Iraqi flying instructors who had been trained at the RAF's Central Flying School in England, joined seconded British RAF instructors who were working at Hinaidi aerodrome. Amongst the new trainee pilots on the Fourth Course or Intake who arrived on 6 August 1936 was Munir Abbas Hilmy. Born in Baghdad in 1909, he would eventually rise to command the Iraqi Air Force. Jalal Jafar al-Awati was six years younger and was only commissioned as a Second Lieutenant on 7 October 1936. He would join the Fifth Course of trainee pilots the following year and would similarly rise to command the Iraqi Air Force.

Before the military coup which shook Iraq on 29 October 1936, the RIrqAF had only 37 pilots and 55 aircraft, while most military expenditure went to the Iraqi Army. There was also ongoing competition between the Army and Air Force Headquarters, both located within the Ministry of Defence building, and it seems clear that the RIrqAF was striving for greater autonomy. On 19 January 1936 the RIrqAF got a new Director, Major General Shakir Abd al-Wahab who had been born in Baghdad in 1889. However, he was replaced less than six months later by *Za'im* (Brigadier) Khalid Mahmud al-Zahawi who had also been born in Baghdad in the same year. Shakir Abd al-Wahab became the Director General of Military Accounts for several years, while Brigadier al-Zahawi would retain his position through the forthcoming coup but would be moved aside on 2 November, thereafter carrying out various duties close to the person of King Ghazi.

By now it was clear that aviation had caught the imagination of Iraq's educated elite, especially amongst certain members of the royal family. Even before coming to the throne on the death of his father King Faisal, Crown Prince Ghazi had often flown with Sabah Ibn Nuri al-Sa'id, the son of Nuri al-Sa'id Pasha who was Prime Minister of Iraq no less than eight times. Sabah was notorious for once flying under the King Faisal I Bridge over the river Tigris in Baghdad. In 1936, during one of those periods when his father was out of office, Sabah Ibn Nuri married a wealthy Egyptian heiress and had two sons, one of whom was named Falah and subsequently flew as King Husain of Jordan's personal pilot.

It was again in 1936 that a certain Ali Muhammad of Baghdad obtained a Westland Widgeon III (c/n WA 1684) which was given the Iraqi civil registration YI-ABC. This small machine was subsequently the first to be owned by the Baghdad Aeroplane Society, though it was either destroyed or demolished in December 1938. Also in 1938, the Baghdad Aeroplane Society obtained three significantly larger De Havilland DH 89A Dragon Rapides registered as YI-FYA, YI-HDA and YI-ZWA. They were later passed to the RIrqAF and some Iraqi officers were photographed with YI-HDA which then had the words "Al-Khatut al-Jawiyah al-Jama'ah al-Tayran al-Iraqiyah" (Airlines of the Iraqi Aviation Association) on its nose. Their serial numbers remain unknown. Nor is there a gap in the sequence where they might have fitted, unless these Dragon Rapides were given something in the range 146 to 150 which had been allocated to six Gloster Gladiators, ordered in October 1940 but never delivered.

By that time, however, the RIrqAF had unfortunately been caught up in the first of the coups which were to plague Iraq. This occurred on 29 October 1936 when Taha al-Hashimi, who had been Chief of Staff of the Iraqi Army since 1923, was out of the

The first 12 Hawker "Iraqi Audaxes", also known as the *Nisrs*, were purchased by the Royal Iraqi Air Force (RIrqAF) in 1934. They included serial number 28 which is seen here at Brooklands with its famous banked motor racing circuit in the background, before it was delivered to Iraq. (Albert Grandolini collection)

De Havilland Dragon number 19 of Royal Iraqi Air Force, apparently after a bad landing. However, the undamaged condition of the propellers suggests that neither engine was running at the time, so perhaps this accident was a result of the machine not being tied down during a high wind. (Albert Grandolini collection)

country in Turkey. Like several other senior Iraqi soldiers, Taha al-Hashimi, had been an officer in the old Ottoman Army, serving in both the First and Second Balkan Wars. He had joined the Arab "Secret Covenant Society" headed by Abd al-Aziz al-Masri, before being sent to Yemen in 1914. There he had fought against British forces around Aden, surrendered to the latter in 1919 and went to Damascus where King Faisal of Syria appointed him Director of Public Security in 1920. Following Faisal's defeat by the French, Taha al-Hashimi travelled to Turkey where he became President of the History Department in the Turkish General Staff headquarters. At last, returning to Iraq in 1922, he was appointed by Faisal, ex-King of Syria and now King of Iraq, as Commander of the Mosul Military Region before becoming Chief of Staff a year later. Meanwhile his older brother, Yasin al-Hashimi, became Prime Minister of Iraq on 17 March 1935 and was still in post at the time of the coup.

Seizing what seemed to be a suitable moment, Bakr Sidqi, the Chief of the Iraqi General Staff and a man whom the British had previously favoured, seized power assisted by Major Muhammad Ali Jawad, the senior staff officer of the RIrqAF and brother-in-law of a subsequent Iraqi military ruler, Abd al-Karim Qassim. King Ghazi, by now King of Iraq, was in his "Palace of Flowers" when, at 8:30 a.m., a Flight of RIrqAF aeroplanes scattered leaflets across the Iraqi capital. Signed by Brigadier Bakr Sidqi who styled himself "Chief of the National Reform Force", these gave a deadline of 11:30 a.m. for the existing government of Yasin al-Hashimi to resign or be dismissed by the King.

In fact, it seems likely that the coup was undertaken with King Ghazi's knowledge and implicit support, and in all probability that of the British Embassy in Baghdad. Perhaps to the coup leaders' surprise and concern, nothing then seemed to happen, so five aeroplanes again appeared over Baghdad and this time dropped four small bombs. One fell close to the Prime Minister's office, a second near the parliament building. According to one version of this event, seven passers-by were killed near the parliament. According to another version the only casualty was an unfortunate coffee shop attendant who was seriously injured near the Council of Ministers building, probably by the bomb which had been aimed at the Prime Minister's office. In fact, the air raid was unnecessary as Prime Minister Yasin al-Hashimi had already submitted his resignation. According to some sources 11 aeroplanes also scattered leaflets over Baghdad, announcing that he had been replaced by the more conservative and anti-reformist Hikmat Sulayman. Nuri al-Sa'id was hurriedly taken out of the country by the British, while Yasin al-Hashimi and Rashid Ali al-Kaylani (Gaylani) made their way to Lebanon as Bakr Sidqi led his troops into the Iraqi capital where a large parade was promptly held.

A few days after the coup, on 3 November 1936, *Qa'id al-Tayar* Muhammad Ali Jawad was appointed Director of the Royal Iraqi Air Force, a position he would hold until his assassination on 11 August 1937. Quite what the officers of the RIrqAF made of the coup is unclear, but later events suggest that even within this small group of young officers, deep divisions of opinion were already developing. The Iraqi coup did, however, reinforce British prejudices about the RIrAF. When it was followed by further coups in which elements of the Iraqi Air Force again took part, a significant number of British military personnel came to the conclusion that "Arabs could not rule themselves", a view which still persists in some western military circles. Meanwhile, Taha al-Hashimi had been replaced as the Iraqi Chief of Staff by the coup leader, Bakr Sidqi, who refused to allow his predecessor to return. In fact, Taha al-Hashimi was only able to do so after Bakr Sidqi was assassinated.

Not entirely to British liking, Bakr Sidqi immediately set about strengthening the Iraqi Army and RIrqAF. He also did so more openly than his predecessor while channelling more funds and munitions to anti-Zionist and anti-British Arab activists in Palestine.

The first Savoia-Marchetti SM 79B built for the Royal Iraqi Air Force, photographed in Italy before it was given RIrAF markings and the serial number 100. (Author's collection)

Savoia-Marchetti SM 79B bombers, purchased for the RIrqAF and wearing Iraqi insignia, lined up at Malpensa aerodrome outside Milan in northern Italy on 8 October 1938, prior to delivery. (Garello archive)

De Havilland DH 82A Tiger Moths of the RIrqAF in formation, photographed while escorting the arrival of SM 79B bombers at the end of their delivery flight from Italy. Iraq acquired 27 Tiger Moth primary trainers and similar DH 83 Fox Moths in 1933 and 1934. (Garello archive)

who would play a significant role in Iraq's acquisition of Italian military aircraft. This was much to the disquiet of the British who continued to regard Iraq as part of their sphere of influence even though the country was not, nor had ever been, part of the British Empire. As early as April 1937 *Qa'id al-Tayar* Jawad was in Europe, seeking more powerful aeroplanes and hoping to diversify Iraq's sources of military supply. Here he expressed his admiration for Italy's fascist form of government – though he might have said this to win a better deal. Jawad also visited aircraft manufacturers in Britain, which eventually resulted in an order for 15 Gloster Gladiator fighters, and in Germany. But it was in Italy that the most dramatic decisions were made, resulting in Iraq acquiring significantly more advanced – and perhaps overly ambitious –

In 1936 *Ra'is (Naqib) Tayar* Akram Mushtaq had handed over the leadership of No. 1 Squadron to *Ra'is (Naqib) Tayar* Mahmud Hindi, though only on a temporary basis. In fact, early in 1937 Hindi was himself replaced by *Ra'is Awal (Ra'id) Tayar* Musa Ali

Savoia-Marchetti SM 79B bombers and Breda Ba 65 ground attack aeroplanes.

In fact, the Iraqi delegation, which consisted of Muhammad Ali Jawad the Commanding Officer of the RIrAF, Hafzi Aziz the CO of No. 2 Squadron, and Jawad Husayn a technical officer, was shown the second prototype SM 79 (c/n 3510) powered by Fiat A 80 RC41 engines at the Savoia-Marchetti factory. Jawad was allowed to take the control of one of these aeroplanes and was so impressed that informal talks promptly started for the purchase of two of the twin-engine SM 79B version of this machine. In its three-engine form as the *Sparviero* "Sparrowhawk", the SM 79 was probably the most successful Italian aeroplane of the Second World War. The Iraqis also wanted an option on purchasing up to ten SM 79Bs but asked if the type could be powered by Bristol Pegasus engines similar to those already in the RIrqAF's Hawker Nisr Audaxes. This was turned down, despite ongoing problems of reliability with the Fiat engines. Similarly, the French Gnome-Rhone 14K Mistral Major, which Jawad suggested, was not available for export.

From Italy, Jawad and the Iraqi delegation went to Germany where the Italians feared they would find suitable aeroplanes immediately available. In the event, nothing came of the German visit, but on 8 June the Italian Foreign Ministry received unconfirmed reports from "a usually reliable source" that the Iraqis had agreed to buy 15 British Gloster Gladiators, this information being passed on to Savoia-Marchetti. On 10 and 11 June 1937, the Iraqi Ambassador to Italy, Muzahim al-Pachachi (al-Bajah Jiy) who would later become his country's prime minister during the Palestine War, and Luigi Capè the Managing Director of Savoia-Marchetti, signed a contract for an Iraqi purchase of five twin-engine Fiat A 80 powered SM 79Bs (serial numbers 99 to 104). One was to be handed over immediately, another in 30 days' time, a third in 90 days, and the remainder between 120 and 150 days later.

This agreement also formalised the Iraqi purchase of 15 single-engine Breda Ba 65s along with 20 Fiat A 80 engines needed for both types. The SM 79Bs would include an early form of autopilot and both types would be supplied with machine guns, radio and,

Hawker *Nisr* or Iraqi Audax number 28 of the RIrqAF in flight, probably over southern England before delivery. At some point Iraq's Audaxes were given additional markings in the form of a diagonal stripe around the fuselage, probably in the Iraqi colour of green, black, white and red. (Albert Grandolini collection)

Hawker Iraqi Audax or *Nisr*, number 28, photographed at Brooklands before the aircraft was delivery to the RIrqAF. In the background is the famous Byfleet Banking at the southern end of Brookland motor racing circuit. (Albert Grandolini collection)

where necessary, photographic equipment. Eventually the deal would include spare parts, parachutes, ammunition and a new all metal Saporiti aircraft hangar of the kind sold to Saudi Arabia. The agreement gave a significant boost to the Italian aero-industry financially and in prestige. Mussolini's government had, for a quarter of a century, been very keen to boost Italy's image across the world (see Volume 3) and the Iraqi deal seemed to open up encouraging possibilities in a Middle East dominated by Italy's rivals, Britain and France.

For Iraq, meanwhile, the apparently potent Breda Ba 65s would form a new squadron, No. 5, established on 25 September 1938 and commanded by *Rais Awal (Raid) Tayar* Hafzi Aziz, a post he held until 1941. Aziz had been sent to Italy early in June to join Jawad to help sort out the details of what was a substantial arms deal. Once assembled, No. 5 Breda Squadron was ready to take to the air on 9 November at Hinaidi, the unit's home. On the Italian side, the Ministry of Defence in Rome chose Regia Aeronautica

The first Savoia-Marchetti SM 79B made for the RIrqAF, now with Iraqi national markings and the serial 100, plus an SM 79B logo below the cockpit. It was photographed at Cameri near Novara in northern Italy in July 1937. Subsequent Iraqi SM 79B would lack the SM 79B logo and have more glazing on their noses. (Garello archive)

Engineer Sergeant Major Bosio, Fiat engineer Mr Ramella, and Mr Maffezzoni of Savoia-Marchetti, an expert in the SM 79B. They would form the technical team sent to Iraq.

While in Italy, Muhammad Ali Jawad trained on a dual control SM 79B with Savoia-Marchetti's instructor, Mr Passaleva. His instruction began on 24 June but after an intensive fortnight, during which Jawad showed himself to be a skilled and attentive pupil, the programme was paused to allow Colonel Tesini of the Italian Air Force's Technical Supervision Office to test the aeroplane prior to its handover. This did not take long and on 6 July Jawad flew two "solos" – meaning he was at the controls from take-off to landing – at the Cameri aerodrome near Novara.

The following day this machine was formally handed over with RIrqAF markings and the serial number 100. Jawad then took off with Engineer Sergeant Bosio and technician Maffezzoni on the first leg of their flight to Iraq. International aviation law meant that the aeroplane was unarmed, the only military equipment aboard being the bombsight. They flew via Belgrade, Istanbul and Aleppo, reaching Baghdad on 9 July without problems after ten hours thirty-five minutes' flying time. Not surprisingly the arrival of the RIrqAF's first really modern bomber was greeted with huge enthusiasm and the usual official receptions.

Jawad was now ordered to show the RIrqAF's powerful new Italian bomber in various parts of the country, resulting in a busy round of demonstration flights beginning on 10 July when the SM 79B went up from both the RIrqAF's main aerodrome of Hinaidi, which was shared with the RAF, and from the "Royal Aerodrome" which presumably meant al-Muthana outside Baghdad. Their purpose was to give other Iraqi officers experience of the SM 79B, its fast rate of climb to 6,000 metres and its speed. On 15 July the machine was formally presented to King Ghazi in the presence of various members of the diplomatic corps in Baghdad though not, apparently, anyone from the Italian embassy. On this occasion two demonstration flights were performed, in at least one of which the King sat alongside Muhammad Ali Jawad in the co-pilot's seat, and was duly impressed.

Unfortunately, things then went dramatically wrong. There had already been rumours concerning sabotage and spying centred upon a can of Castrol lubricating oil. Its seal had been broken and there were traces of silicon carbide (carborundum) abrasive powder. After this discovery Jawad had been advised only to use oil from cans with unbroken seals. He had also told the RAF's No. 55 Squadron to leave Hinaidi air base, which was now under Iraqi command, and transfer its Vickers Vincents to the British air base at Dhibban (Habbaniya). The British were not used to taking orders from Iraqis and No. 55 Squadron's move would not be completed until September 1937.

Two days after demonstrating the SM 79B to King Ghazi, the aeroplane took off again early in the morning of 17 July. Everything seemed to be working well until the Savoia-Marchetti came in to land, whereupon the engines did not respond properly and the brakes failed to function as they should. The machine rolled on and on until it came to a halt behind a hangar. Immediate investigations found evidence of "tampering" to both the throttle and braking system. Jawad, however, recommended that these findings be kept under wraps because he did not want to point a public finger at those he already suspected – namely the British.

Two days later General Bakr Sidqi, the leader of the recent coup and now the political master of Iraq under King Ghazi's somewhat nominal rule, chaired a meeting of almost 90 senior military officers. During this meeting Jawad took up the SM 79B for a demonstration flight with a dozen Iraqi officers on-board. Everything was fine, so Jawad went up again with another group of officers plus the Italian specialists Bosio and Maffezzoni. On landing at a speed of between 115 and 120 kph the undercarriage collapsed and the machine skidded across the aerodrome in a cloud of dust. Fortunately, no one on-board was injured apart from Bosio who suffered bruising.

This accident was probably the result of an error by the crew, perhaps accentuated by the SM 79B being overloaded with passengers and having been fully refuelled before the second flight. The propellers and undercarriage of course needed replacing, in addition to which there had been minor damage to the fuselage and wings. This event caused gloom and disappointment throughout the RIrqAF, whereas the British naturally presented it as evidence of the inferiority and unsuitability of Italian aeroplanes. The RAF offered to help salvage the SM 79B because they had the necessary heavy lifting equipment, but the crane's brake failed and the large twin-engine aeroplane was dropped from a height of two metres. This happened again on a second attempt, resulting in such severe damage that the bomber needed to be dismantled for major repairs.

A very potent addition to the RIrqAF's inventory in 1938 were 15 Breda Ba 65 ground attack aircraft. Lined up here before delivery to Iraq, they include two Breda Ba 65B dual control instructional aircraft which lacked a dorsal gunner's turret. (Ahmad Sadik collection)

Iraqi prestige was now at stake, so Jawad insisted that all repair work must be done by the RIrqAF alone. Nevertheless, this proved impossible and in reality the repairs could only be done in Italy, so the aeroplane was dismantled. Its engines were removed, the fuselage and wings being crated and sent back to Europe by sea. Even the traditionally reserved language of official British governments could not hide their delight at this outcome. For a while the Italian engineers at Hinaidi continued to instruct their Iraqi colleagues on the Fiat A.80 engines which had been removed from the damaged SM 79B, until these were also sent back to Italy.

British and Iraqi representatives signing an agreement on 2 June 1937 whereby Mosul aerodrome was transferred from the RAF to the RIrqAF. The British are trilby hat wearing civilians whereas the Iraqis are uniformed military men, including Musa Ali, Ahmad Hamdi Abd al-Jalil Husain and one other. The air base was then renamed Mu'askar al-Firnas (Al-Firnas Aerodrome) after the Arab-Andalusian scholar Abbas Ibn al-Firnas who flew with some degree of success in the 9th century (see Volume One). (Iraqi Air Force archive)

With almost nothing to do, the Italian technicians found that they were no longer paid regularly by the Iraqi government. There was, meanwhile, considerable annoyance in Italy where it was felt that this accident had harmed the reputation of the Italian aero-industry. So, Count Galeazzo Ciano, Mussolini's Foreign Minister, urged General Giuseppe Valle the Chief of Staff of the Regia Aeronautica to accelerate the delivery of a second SM 79B to Iraq. Unfortunately, Savoia-Marchetti were currently concentrating on getting five SM 79CSs ready for the forthcoming Paris-Damascus-Istres air race on 20 August 1937. There was also some question about the viability of repairing the Iraqi SM 79B and whether it might be simpler for the Iraqis to achieve their required number of four bombers by purchasing an additional machine.

Meanwhile Iraq's acquisition of 15 Breda Ba 65 ground attack aeroplanes, powered by Fiat A 80 RC 41 engines, appeared to be progressing more smoothly. The sale had been negotiated on 10 and 11 June 1937 between the Iraqi government and a consortium of Italian companies called Aerocons which had recently been formed to reduce undue competition between Italian companies for overseas orders. It was then agreed that the price be paid in two stages on delivery of the machine. The first would consist of one standard machine and two dual control versions by the end of August 1937, with the remaining aircraft delivered in September, all for a newly established No. 5 Squadron RIrqAF. This contract included 25 Fiat A 80 engines, thus providing ten spares. Jawad had wanted the Bredas to be powered by French Gnome-Rhone K14 engines or Italian Isotta Fraschini engines, but the Italian government insisted upon the Fiats in order to give support to that company – largely for political reasons. Iraqi Breda Ba 65s were of the two-seater version and, apart from the two dual control machines, had a Breda Type L dorsal turret armed with a single 7.7mm machine gun.

During the first half of 1937 all military aerodromes in Iraq, with the exception of Habbaniya west of Baghdad and Shaibah in the far south, were handed over to Iraqi control as a result of a new treaty between the UK and Iraq. In March the British vacated Hinaidi, which the RIrqAF renamed Mu'askar al-Rashid (Rashid aerodrome, after the famous Abbasid Caliph Harun al-Rashid). It thus became home to the bulk of the Iraqi Air Force. This was followed by the handover of Kirkuk aerodrome, which the Iraqis renamed *Mu'askar al-Hurriyah* (Freedom Airfield). On 2 June the RAF also handed over Mosul aerodrome, which the Iraqis renamed Mu'askar al-Firnas, after the remarkable Arab-Andalusian scholar Abbas Ibn al-Firnas who had flown with a degree of success in the 9th century and had survived the event (see Volume 1).

Italian Lieutenant Bertotto wearing a solar topee, with several RIrqAF officers standing in front of one of Iraq's new Savoia-Marchetti SM 79B bombers in Iraq. (Garello archive)

Habbaniya and Shaibah remained under British control because they were regarded as essential for imperial communications between the UK, India and further east. On 2 February 1937 the ground personnel of No. 1 Squadron RIrqAF left Baghdad for Mosul and four days later Mosul aerodrome became this unit's official home at a formal ceremony during which the British flag was lowered and that of Iraq raised. This necessitated a new set of regulations, being the first time a RIrqAF squadron had been permanently based away from the capital. It came in the form of Army Chiefs of Staff Regulations Book No. 439, dated 21 February 1937 which stated that No. 1 Squadron was considered an independent unit, linked to the Army's Mosul Regional Command and taking its orders from that command. Its administrative staff were responsible for paying salaries, purchasing fuel and paying the taxes associated with such purchases, but had to send regular reports and accounts to RIrqAF HQ in Baghdad. No. 1 Squadron's administrative staff similarly found themselves responsible for hiring and firing civilian technicians and other workers, though such actions still had to be cleared with Baghdad. Thus, the Mosul air station almost became a small Air Force in its own right.

Meanwhile the so-called Euphrates Uprising rumbled on in the south of the country. This time it seems that Nos. 1 and 3 Squadrons joined forces and when a Hawker Nisr was shot down by hostile groundfire in the area of al-Samawah on 13 June, killing both the crew. The pilot *Mulazim Awal Tayar* Anwar Mustafa was from No, 1 Squadron, while the observer, *Jundi Aslahah* (Weapons Soldier) Muhammad Ibrahim, was from No. 3 Squadron.

Then came the Iraqi coup – or more correctly "counter-coup" – of 11 August 1937. The previous coup of 1936 had resulted in General Bakr Sidqi becoming de facto ruler of Iraq, though he was never formally a member of the government and instead remained commander of the country's armed forces under King Ghazi. General Sidqi and *Qa'id al-Tayar* Muhamad Ali Jawad, the Director of the RIrqAF and CO of No. 2 Squadron, were on their way to Turkey on an official visit when they were assassinated in the garden of Mosul aerodrome (Mu'askar al-Firnas) by a soldier named Muhammad Ali Tal Afari. The latter was not apparently acting alone, and a small group of relatively senior Iraqi officers who were suspected of planning the assassinations were understood to have included Mahmud Salman al-Janabi. He himself would command the RIrqAF from 11 June 1938 until 30 May 1941 and certainly did not turn out to be pro-British.

Both Sidqi and Jawad were buried with full military honours in Baghdad the next day, and their deaths led to the return to power of what has been called the "pro-British faction". Hikmat Sulayman was ousted as prime minister to be replaced by Jamil al-Midfa'i, who remained in post until 25 December 1938. Four days after Jawad's assassination, his place at the head of the RIrqAF was taken by an Army officer, Salah al-Din Ali al-Sabbagh. Born in Mosul in 1899, he seems to have been selected as a temporary leader at a moment of potential crisis, and only retained this position until 11 October 1937.

In Rome, Bakr Sidqi and Muhamad Ali Jawad had been seen as friends of Italy and their killing caused concern in the Italian government and military. This deepened when al-Sabbagh was replaced by the clearly pro-British *Ra'is Tayar* Akram Mushtaq. He promptly tried to persuade the new Iraqi government to cancel existing contracts for Italian military supplies, or at least to replace the SM 79B with other equipment such as anti-aircraft guns and Ansaldo tanks (probably L3/35 tankettes). The Italian companies and government resisted vigorously and so things remained unchanged; indeed, in March 1938 Iraq confirmed an order for four new Savoia-Marchetti bombers, the complete repair of the first SM 79B, plus the services of Italian aircrew and technicians to train the RIrqAF's new No. 6 Squadron in the operation of these aeroplanes.

The prestige of the Savoia-Marchetti SM 79 had meanwhile revived, not least as a result of the Paris-Damascus-Istres air race of 20 August 1937 where five of these machines came first, second, third, sixth and seventh. The type was also far superior to anything that the British RAF currently had in service in the Middle East. On 15 September the Iraqi government, which would remain chronically short of money until the later development of the country's oil industry, requested that part of its payment be made in the form of a quarter of a million lire worth of high-quality dates, which the Italians declined. The Italian Air Force, and the companies involved, were also reticent about sending their technical personnel to Iraq, anticipating trouble with the British. Nevertheless, the Italian government put its foot down, stating that if these men were not sent, 'RAF pilots would end up being instructors on our aircraft'.

Before the SM 79Bs were delivered to the Iraqis, some modifications were made to the dorsal gun position and to the glazed nose. The four new machines, with Iraqi serial numbers 101 to 104, were tested by Italian civilian pilot Passaleva and *Tenente* (Lieutenant) Charles Bertotto who had been appointed as instructor for the Iraqi pilots. In addition to speaking excellent English, which would be necessary for proper communication with his Iraqi colleagues and students, *Tenente* Charles Bertotto already had combat experience during the Italian invasion of Abyssinia (Ethiopia) and Italy's involvement in the Spanish Civil War.

Delivery of the SM 79Bs was nevertheless further delayed because the Italians feared that the Iraqi government would pull out of the deal at the last moment. This time the British chose to be more helpful in the wake of a diplomatic agreement with Italy whereby both promised not to interfere in the others' "sphere of influence" within the Arab world – a little-known act of appeasement in the run-up to the Second World War. The British therefore put forward a certain Mr Selmi Zibuni as a mediator between the now rather acrimonious Iraqi and Italian parties.

In March 1938 Iraqi Prime Minister Jamil al-Midfa'i confirmed his country's order for these aircraft, having insisted that during the delivery flight each machine would have an Iraqi co-pilot aboard, and that the Italian consortium would cover the costs of delivery flights though this was reduced from 70,000 to 50,000 lire per aeroplane. The Italians were also obliged to send two pilots and five technicians to train Iraqi personnel, these men to remain in Iraq for three months. They would not receive payment from the Iraqi government but would be given food and accommodation by the Iraqi armed services. Italy would cover the costs of any necessary repairs during the delivery flight and the damaged first SM 79B would be repaired and delivered in due course.

Having given way on a number of points during these negotiations, the Italians expected to soon receive further Iraqi orders for aeroplanes, engines, machine guns for the aircraft, and ammunition (both 7.7mm and 12.7mm) while hoping this sale would stimulate orders from other parts of the world. On 20 July 1938 the Aerocons consortium finally declared the four new SM 79Bs to be ready and asked the Regia Aeronautica to provide crews to fly them to the Middle East. Those chosen were all from Italian Air Force bomber units, some having experience of combat in Ethiopia and Spain.

Meanwhile, the RIrqAF's attempts to get the RAF to remove all its facilities from Hinaidi to Dhibban had not yet succeeded. Hinaidi would be needed as a training centre for the SM 79Bs and Breda Ba 65s, but in September 1937 another RAF unit, No. 70 Transport Squadron with its Vickers Valentias, moved in temporarily before going to RAF Dhibban (Habbaniya) and then to Egypt. Despite the existence of a pro-British government in Baghdad, relations between the RIrqAF and RAF were already sensitive if not yet tense, though this did not stop the purchase of 15 ex-RAF Gloster Gladiators by the Iraqis.

According to Italian intelligence reports, these Gladiators cost the Iraqis the equivalent of 10,000 lire each and, according to a British Foreign Office document dated 31 December 1937, the Iraqi Ministry of Defence was proposing to form a new fighter squadron with Gladiators "shortly". As part of the process two RAF NCOs, Sergeants Crane and Wood, had been selected for service with the RIrqAF. They were expected to be attached for three years, both having recently 're-engaged to complete [their] time for pensions', and would replace Sergeants Elvin and Fraser who were currently with the RIrqAF. In the event, the Second World War interrupted this plan.

The first batch of nine Iraqi Gladiators (serial numbers 79 to 87) was delivered to Basra aboard the British steamer SS *Clydewater* in mid-November 1937. Assembled and tested by RAF personnel at the nearby British airbase, they were then handed over to the RIrqAF and, on 1 December, a Gladiator squadron, No. 4, was at last established under Captain Hafzi Aziz. He would remain in command until 1938, when he was sent to Italy as part of the team to prepare for the RIrqAF's acquisition of Italian aeroplanes. After this he was placed in command of the new No. 5 Squadron whose SM 79 bombers were based at Mu'askar al-Rashid outside Baghdad. The aeroplanes arrived packed in crates and were assembled by Iraqi technicians at Hinaidi (Mu'askar al-Rashid) where No. 4 Squadron and its Gladiators would remain until 1940, seven more being delivered during 1938 (serial numbers 88 to 94). Meanwhile the year 1937 saw the expansion of Iraqi Air Force pilots to 127 as the result of an intensive recruitment and training programme.

In August 1938 the men selected to bring home the SM 79Bs arrived in Italy. They were: *Rais Awal (Raid) Tayar* Hafzi Aziz the CO of the group; *Rais Awal (Raid) Tayar* Bahgat Ra'uf; *Ra'is Tayar* (Captain Pilot) Naqib (almost certainly Majd al-Din Abd al-Rahman al-Naqib); *Ra'is Tayar* (Captain Pilot) Abadi (probably Abd al-Kadhim Shaikh Abadi); and *Mulazim* (Lieutenant) Yahya (perhaps Yahya Ithniyani, who was one of the Iraqi officers who had recently learned to fly at RAF College Cranwell (see Chapter 5)). As agreed by both governments, they were confined to Malpensa aerodrome near Milan, apart from short trips to Turin and the Alpine resort of Sesto, undertaking an intensive training programme which lasted until the end of September.

For the flight to Iraq, these RIrqAF aeroplanes had to be given nominal civilian registrations, though they must have been applied in very small or very indistinct paint as they are not visible in photographs taken during the delivery flight. In case they needed to summon assistance, the machines were equipped with more powerful radios than the original contracts envisaged. An experienced radio operator named Mercalli was also supplied by the Regia Aeronautica's 10th BT Bomber Wing.

It was not until 8 October 1938 that RIrqAF SM 79B, serial number 101, took off from Malpensa outside Milan in northern Italy at 11:26 a.m. and headed for Baghdad. At the controls was *Tenente* Charles Bertotto, accompanied by *Qa'id* Hafzi Aziz and Italian NCOs Mercalli and Scapparone. They were closely followed by SM 79B number 102 flown by *Tenente* Niggi with *Ra'is Tayar* al-Naqib and *Serg*ente Cozzi, number 103 flown by *Tenente* Persico with *Ra'is Tayar* Abadi, *Ra'is Tayar* Abd al-Fattah and Mr Ramella of the Fiat company, and number 104 flown by *Tenente* Magri with *Mulazim* Yahya and Ugo Cheloni, a mechanic who would join the training mission.

Flying in a loose formation which extended over four kilometres, their route took them from Malpensa, via Catania,

One of the first Italian-made Savoia-Marchetti SM 79B bombers sold to Iraq, seen here at Malpensa aerodrome outside Milan in the summer 1938. The Iraqi pilots and their Italian instructors are preparing for a familiarisation flight. (Garello collection)

Flight engineer Ugo Cheloni with a Royal Iraqi Air Force SM 79B in Iraq, probably at Mu'askar al-Rashid aerodrome outside Baghdad. Ugo Cheloni was a highly skilled mechanical engineer who joined the Italian training mission to Iraq. (Garello archive)

One of the RIrqAF's newly delivered Savoia-Marchetti SM 79B bombers (number 103) being shown to King Ghazi, probably in October 1938. (Garello archive)

Tripoli, Benghazi, Umm Sa'ad and Cairo-Heliopolis to Baghdad. On 11 October the formation took off from Benghazi at 2:40 p.m. but as they neared Darnah (Derna) after about one hour's flying, *Tenente* Bertotto reported the failure of one of his fuel pumps and so the formation returned to Benghazi, the following day being spent repairing number 101 and sealing a leak in one of number 102's radiators. On the 13th the flight resumed, landing at Umm Sa'ad close to the Egyptian frontier at 10:45 a.m. Unfortunately, the runway at this forward aerodrome was short and strewn with stones. *Tenente* Persico in number 103 had to yaw slightly as he ran off the end of the runway, damaging one of his tyres. The Iraqis were nevertheless impressed by the fact that this sturdy machine suffered no other damage. Back in the air at 2:45 p.m., the formation crossed into Egyptian airspace at 4,500 metres altitude, reaching Heliopolis aerodrome at 5:30 p.m. On the ground the RAF was slow to switch on the landing lights, so only the fourth SM 79B had their benefit. The first three descended as rapidly as they could in the gathering dusk, earning the somewhat grudging admiration of British observers on the ground.

A "day of rest" in Cairo was spent checking the aircraft and their engines before the final leg of the journey. On 15 October 1938 they all took off and climbed to 4,500 metres before heading east. Over the Dead Sea storms were encountered and the aeroplanes descended to lower altitude. After crossing the Iraqi frontier, it seems that a stop had been planned at the small airfield at Rutbah but sandstorms persuaded Bertotto to press on for Baghdad because they had sufficient fuel. As the bombers flew above the river Euphrates and approached the Iraqi capital, the formation pulled closer together and a number of fine air-to-air photographs were taken. At 1:15 p.m. they landed outside Baghdad, to be greeted by the RIrqAF's commanding officer, *Ra'is Tayar Naqib* Akram Mushtaq and Crown Prince Talal Ibn Abdullah of Transjordan. A photograph showing an Iraqi SM 79B escorted by two RIrqAF Gladiators is said to have been taken at the end of the delivery flight, though this cannot be confirmed.

As soon as these ceremonies were ended, Bertotto and Cheloni started the training programme for Iraqi air and ground crew, other Italians returning home by 21 October. The Iraqi pilots, having been used to older and much slower single-engined biplanes, at first found conversion to the demanding Savoia-Marchetti difficult. Between December 1938 and January 1939 Italian records show that the following Iraqi officers were trained in SM 79Bs (unfortunately some of the names are only known through their Italian form): *Ra'is Tayar* Ibrahim Jawad (not, of course, the Jawad who had been assassinated at Mosul during the coup of 11 August 1937 and not the more junior *Mulazim* Jawad Husayn who flew Breda Ba 65s during the Anglo-Iraqi conflict of 1941); *Ra'is Tayar* Majd al-Din Abd al-Rahman; *Ra'is Awal Tayar* Ra'uf, *Mulazim* (Lieutenant) "Cherim" [Sharim?]; *Mulazim* (Lieutenant) Issam; *Mulazim* (Lieutenant) Hamid; *Mulazim* (Lieutenant) Rashid; *Mulazim* (Lieutenant) Rushad; and *Mulazim* (Lieutenant) "Sciar" [Shiar?].

After six months this intensive training resulted in 15 Iraqis being qualified to fly the twin-engine aeroplanes as "first pilots", including the senior men of No. 2 Squadron, *Qa'id* Hafzi Aziz and *Ra'is Tayar* Majd al-Din al-Naqib. The SM 79B squadron had also been given its own administrative office in the RIrqAF headquarters in Baghdad, headed by *Ra'is Tayar* Abd al-Fattah who developed a close working relationship with the Italian *Tenente* Charles Bertotto. They were particularly keen that the SM 79B crews became competent in astronavigation which had previously been neglected in the RIrqAF because the Iraqis did so little night flying. There was also the fact that no Iraqi pilot was experienced enough for the role of instructor on the SM 79B. The remaining four SM 79Bs arrived in Iraq on 23 March 1939 but No. 6 Squadron was not formally established at Hinaidi aerodrome until 31 May when *Ra'is Tayar* Ibrahim Jawad took over command, a position he would hold until 30 May 1941.

Maintaining and operating such advanced machines was a daunting task for the RIrqAF which was not helped by the fact that wind-blown sand easily penetrated the bomber squadron's

The Iraqi Military Mission visiting Rome in 1939. From right to left they are Hafzi Aziz, an unnamed pilot, Ibrahim Hamdi al-Rawi the mission leader, Italian Col. Paolo Sbernadori, Ibrahim Jawad and Abd al-Rahman the Iraqi ADC. (Garello collection)

Breda Ba 65, serial number 118 (c/n 64097) was built for the RIrqAF but suffered slight damage after a wheels up emergency landing at Bresso aerodrome outside Milan prior to delivery to Iraq. Another (c/n 64084) would be written off, though the Italian test pilot escaped unhurt. (Garello archive)

The RIrqAF's Breda Ba 65B number 106 was one of the dual control instructional versions of this ground attack aircraft, which lacked the dorsal gunner's turret. (Garello archive)

79Bs airworthy. In practice, these Iraqi maintenance crews still needed to have one Italian with them as they worked.

Despite such difficulties, Bertotto could report back to Italy that No. 6 Squadron was delighted with the high speed, aerobatic agility at high speed and graceful appearance of its Savoia-Marchettis. The only negatives were its poor manoeuvrability below 220kph, its high stalling speed, the complicated procedures needed when landing, its relatively small bombload and the extreme difficulty of retaining control if an engine failed. Meanwhile, the Iraqi government was determined that No. 6 Squadron should become a potent factor in the turbulent Middle East and, with this in mind, purchased four thousand 2kg bombs and one thousand 15kg bombs, plus 20,000 rounds of ammunition for each of the aeroplanes' guns.

Nor was the delivery of Breda Ba 65s without incident, even before these aeroplanes were handed over. One of them (c/n 64097) suffered minor damage at Bresso aerodrome while another (c/n 64084) was written off on 2 March 1938 when it crash-landed in a field following engine failure, though the pilot Ildebrando Artigiani was unhurt. This machine was replaced by another from a batch being built for the Italian Air Force by the Caproni company at Vizzola Ticino. In September, 15 machines (serial numbers 105 to 119 were delivered to Iraq. Number 106 may have been a replacement for the aeroplane which had crashed on 2nd March.

Unlike the larger SM 79Bs, the Bredas were sent to Iraq by sea, the first ten arriving at Basra on 26 March 1938 after a 13-day voyage. Moving them to Hinaidi air base outside Baghdad took a further month because the Tigris was in flood. A second batch of five reached Baghdad on 30 June, to be given serial numbers following those of the SM 79Bs. Meanwhile the Breda company sent a technical team consisting of pilot *Tenente* Rodolfo Guza, a fitter, a flight engineer and an armourer who had already started assembling the first three Bredas, one of which was a dual control version. Nevertheless, the assembling of the remaining Breda Ba 65s was frequently interrupted by the need to train Iraqi technicians at the same time.

In July, Guza made several demonstration flights, including a private viewing for King Ghazi and his court because the Iraqi monarch had expressed a particular interest in the Breda. *Tenente*

hangar. The RIrqAF's acute shortage of suitably trained technical maintenance crews remained a persistent problem, only four engine specialists, four airframe specialists and three armourers being available to keep No. 6 Squadron's four operational SM

Breda Ba 65s of the RIrqAF lined up at Bresso aerodrome in northern Italy, prior to their delivery to Iraq. Visible from the left are serial numbers 110, 109, 112, 113 and 114. (Garello archive)

Guza also made training flights with RIrqAF pilots, five of the latter being qualified in November 1938 after 37 hours on a dual control machine. Guza reported that they had shown themselves serious and skilled in the low-level flying for which the Breda Ba 65 was designed. It has been suggested that the RIrqAF officer Musa Ali, who had been involved with the purchase or working up of Iraq's Italian warplanes, then advised the Saudi Arabian government in its negotiations to buy Italian Ca.111 trainers and Ca.133 bombers in 1938–1939 (see Chapter 3).

The Iraqi government of course continued its purchase of more old-fashioned British aeroplanes, and to have RIrqAF personnel trained by the British in both Iraq and in the United Kingdom. Apart from political considerations, one reason why Britain was unwilling to supply Iraq with advanced warplanes was the fact that the RAF was currently in the midst of an urgent rearmament programme. A second world war was on the horizon, though politicians and others in Britain and France still desperately hoped that it could be avoided, and as a result mixed messages were sent to the Iraqi authorities. One example, dated 10 October 1938, informed the latter that 'HMG [His Majesty's Government] would welcome the development of the Iraqi forces on lines which would enable them, acting in conjunction with the reduced strength of the British Air Forces, to meet with confidence all the initial requirements of any attack upon Iraq by her neighbours'. Quite which neighbours was not specified.

King Ghazi, who had shown himself to be enthusiastic about aviation, was in favour of the purchase of British equipment. Indeed, he bought a Miles Hawk Trainer Mk. III (c/n 797) for his own use. Named *Shatt al-Arab*, it was given the Iraqi civil registration YI-GFH on 15 August 1938. All the other civil registered aeroplanes owned by King Ghazi were similarly British:

- DH 90A Dragonfly (c/n 7502) registered YI-HMK on 28 April 1936 (subsequently transferred to RIrqAF)
- Percival P.10 C Vega Gull (c/n K.85) registered YI-CPF on 29 October 1937 (subsequently to RIrqAF)
- Percival P.16A Q-6 (c/n Q.22) registered YI-ROH or YI-RQH on 1 April 1938
- Spartan 7W Executive (c/n 7W-19) registered YI-SOF on 8 May 1939 (actually after King Ghazi's sudden death on 4 April 1939)

Meanwhile, No. 2 Squadron reportedly received three De Havilland Rapide aeroplanes on 12 October 1938, though these do not seem to have been given military serial numbers. If such an acquisition is true, it must have complicated the unit's servicing programme as it already operated DH 80 Puss Moths and DH 84

In August 1938 King Ghazi of Iraq purchased a British Miles M.14 Magister, in which he hoped to learn to fly. He named the aeroplane *Shatt al-Arab* after the vital waterway which links the Tigris and Euphrates rivers to the Persian Gulf. (Albert Grandolini collection)

Dragons. However, it is more likely that these were the Dragon Rapides belonging to the Baghdad Aeroplane Society (registered YI-FYA, YI-HDA and YI-ZWA), subsequently operated by the "Airlines of the Iraqi Aviation Association" perhaps in collaboration with the RIrqAF.

Another important aspect of the relationship between the RIrqAF and the British was the secondment of RAF personnel to assist the Iraqis. Flight Sergeant J.T. Darling was already with the RIrqAF but was due to be replaced by Flight Sergeant H.C. Smith after May 1938. Two additional men were also needed but the RAF had difficulty finding candidates, though two suitable NCOs already in Iraq had volunteered: Sergeant B.W. Carter a fitter, and Sergeant T.L. David a metal rigger. According to British official documentation dated 14 April 1938, they could not be released from duties at RAF Dhibban (Habbaniya) and the resulting correspondence went to and fro for a long time.

Such messages made it clear that men seconded to the RIrqAF tended to be those with good records, high experience, "very good character", sometimes single but mostly married, and medically fit for such service. Their trade proficiency was sometimes "exceptional" or "superior". They would receive increased pay but had to wear Iraqi uniform and the great majority already had experience serving with the RAF in various other parts of the Middle East.

Away from the political hothouse of Baghdad and the international rivalries which surrounded Iraq's purchase of Italian aircraft, other parts of the RIrqAF got on with training and normal duties. In 1938 Captain Hamid Ammar took over command of No. 4 Squadron. However, No. 3 Squadron suffered a further fatality on 11 April 1938 when a Hawker Nisr crashed into another aeroplane while taking off from Hinaidi (Mu'askar al-Rashid), *Mulazim Tayyar* Victor Alakah being killed and the other pilot injured. During 1938 the British also supplied Iraq with further, presumably replacement, Hawker Nisrs (serial numbers 95 to 98).

Throughout 1938 a group of senior Iraqi officers attempted to dominate their country's government which was under the experienced and staunchly pro-monarchist Prime Minister Jamil al-Midfa'i. He was in office from 17 August until 26 December 1938, but his cautious and conciliatory policies frustrated some senior military men, notably Lieutenant Colonel Salah al-Din Ali al-Sabbagh who had briefly headed the RIrqAF in 1937 following the assassination of Bakr Sidqi. These officers, known at the time as the "Circle of Seven" but later, having effectively purged their own ranks, as the "Golden Square", currently enjoyed the support of a highly experienced Iraqi politician, Nuri al-Sa'id who had returned from exile in Cairo.

Under pressure from both the Colonels (who were muttering about the possibility of another coup) and Nuri al-Sa'id, King Ghazi eventually bowed to pressure and replaced Prime Minister Jamil al-Midfa'i with the wily Nuri al-Sa'id. Nevertheless, the latter soon came to be seen as excessively pro-British by a significant part of the Iraqi military and much of the broader population. Some of the "Circle of Seven" were undoubtedly anti-British, at least in the sense of being Arab nationalists who wanted the removal of all foreign powers from the Middle East. Lieutenant Colonel Salah al-Din Ali al-Sabbagh and Lieutenant Colonel Mahmud Salman al-Janabi of the RIrqAF were both in this political camp.

Meanwhile the RIrqAF trained on its new Italian and rather less new British aeroplanes. *Ra'is Tayyar* (Captain Pilot) Akram Mushtaq had been both CO of No 2 Squadron and CO of Hinaidi aerodrome as well as being the RIrqAF's senior administrative officer. Such a multiplicity of tasks was characteristic of many small new air forces during this period. In October 1937 Mushtaq had become the RIrqAF's commanding officer but was replaced by Mahmud Salman al-Janabi on 11 June 1938, perhaps partly because he was seen as too pro-British. Presumably in compensation he now received the temporary rank of colonel and remained a senior staff officer. The British were far from pleased.

Mahmud Salman al-Janabi had been born in Baghdad in 1898, graduated from the Military College in Constantinople (Istanbul) on 4 July 1916 and had fought as a junior officer in the Ottoman Army during the First World War. He then served in the Hashemite Arab Army in Syria, after which he had undergone further training, or retraining, in Iraq during 1924 before serving as a cavalry officer in the Iraqi Army from 1925. Subsequently Mahmud al-Janabi served in the Iraqi Royal Guard before commanding the Iraqi

At one point *Al-Aqid al-Tayyar* Akram Mushtaq had been CO of No. 1 Squadron, No. 2 Squadron and Hinaidi aerodrome, as well as being the RIrqAF's senior administrative officer. This highly experienced officer then served as Commander of the RIrqAF from 11 October 1937 and was soon urging his government to cancel further orders for Italian aircraft. (Iraqi Air Force archive)

Hashemite Arab Cavalry School. He served as an aide-de-camp to both King Faisal and King Ghazi, transferred to the new Armoured Corps, was promoted to the rank of Lieutenant Colonel on 8 September 1937 and then qualified as a pilot.

Other officers who subsequently rose to senior rank could be found in a variety of positions in the RIrqAF at this time. They included Sami Abd al-Fattah who, born in Mosul in 1905, graduated from the British Army's prestigious Sandhurst Military Academy as a Second Lieutenant on 1 January 1928 and qualified as a pilot in Britain in May 1931. He then went to the Staff College in Iraq and became a senior Staff Officer on 12 March 1938. Abd al-Fattah would subsequently command the Air Force in the aftermath of the Anglo-Iraqi War of 1941 (see Volume 6).

Another future leader of the RIrqAF was Jissam Muhammad al-Shahir. Born in Baghdad in 1914, he volunteered as one of the Iraqi Air Force's first technicians in 1928 before the RIrqAF was officially created. Initially working as a civilian mechanic, he joined the military in 1932 and was promoted to the rank of corporal. Subsequently volunteering to be trained as a pilot, Jissam al-Shahir earned his wings in 1938 and was later sent to the RAF College of Aviation Engineering in England before being commissioned as an officer. He eventually retired with the rank of Air Brigadier in 1959. Recalled to command the Iraqi Air Force from 1966 to 1968, Jissam Muhammad al-Shahir was a "rare bird", rising from the rank of corporal to command his country's Air Force. Another future commander was *Ra'is Awal (Ra'id) Tayyar* Nasir Hussain al-Janabi who took over the leadership of the No. 1 Squadron from 1938 to 1939. Whether he was related to the RIrqAF's current commander, Mahmud Salman al-Janabi, is unclear.

Although most attention was focussed on the SM 79B bombers of No. 6 Squadron, the men of No. 5 Squadron continued to train on their Bredas. In January three more Iraqis completed their training without serious incident and on 27 February three Breda Ba 65s flew in formation over Baghdad to celebrate an official visit by Crown Prince Reza Pahlevi of Persia (Iran). In fact, Rodolfo Guza the Italian flying instructor remained in Iraq for 15 months, well beyond what was originally agreed, only returning

Retrieving an RIrqAF Breda Ba 65 (number 117) following an emergency landing in Iraq, either with the wheels up or during which the undercarriage failed. This apparently happened on several occasions. (Garello archive)

Breda Ba 65, serial 108 of the RIrqAF, with Italian instructors Carlo Bertotto and Rodolfo Guza, probably photographed in Iraq shortly before the outbreak of the Second World War. (Garello archive)

to Italy briefly to bring back the repaired SM 79B (number 100). Guza's tireless work and tact made him popular amongst his Iraqi colleagues. He also struck up an unexpected friendship with Wing Commander Cullay, the head of the RAF Advisory Mission in Iraq. On one occasion Cullay went up with Guza in one of the Iraqi Breda Ba 65s but had to make an emergency landing when the engine failed – both men emerging unscathed. On another occasion the RIrqAF asked them to carry out a mock combat between a Breda Ba 65 and a Gloster Gladiator, which the Breda was declared to have won, despite the Gladiator's greater manoeuvrability.

This happy state of affairs was nevertheless blighted by gathering war clouds and an accident when a Breda Ba 65 stalled at low altitude. In the event Guza went home in May 1940, on the eve of Italy's entry into the Second World War as Germany's ally. Subsequently, he fought with distinction and became one of the Regia Aeronautica's most renowned bomber and torpedo-bomber pilots. Meanwhile *Ra'is Tayar* Akram Mushtaq, the RIrqAF's senior staff officer, had asked *Tenente* Charles Bertotto to extend his time in Iraq to continue training SM 79B crews. This was permitted by the Italian government, but only for three months.

Much less publicity was given to the British aeroplanes which the RIrqAF obtained during the year before the outbreak of the Second World War. These included 15 Gloster Gladiator Mk. I biplane fighters and some Avro Anson light bombers which, like the Gladiators, were transferred from British RAF stocks. The Gladiator nevertheless remained a demanding machine, and five had been written off by 1940 (numbers 80, 82, 86, 90 and 91), the British somewhat reluctantly supplying one (serial number unknown) as a partial replacement in October 1940. One De Havilland DH 90A Dragonfly light transport was also acquired, having belonged to the late King Ghazi with the civil registration YI-OSD. However, no further mention was made of the royal Percival P.10 C Vega Gull (YI-CPF) which is similarly said to have been transferred to the RIrqAF.

Having been replaced by Mahmud Salman al-Janabi as the RIrqAF's commanding officer in June 1938, Akram Mushtaq remained a senior staff officer before retiring from the Air Force on 9 April 1939 and was appointed Director of Iraqi Civil Aviation nine days later. Mushtaq's temporary rank of colonel was also taken away on 31 May. These were the most obvious signs of increasing anti-British sentiment in the upper echelons of the RIrqAF, but there were others. With war clouds gathering over Europe, and despite such worrying trends, Britain was eager to put its tense relationship with Iraq on a more stable footing. The result was a Draft Joint Iraqi-British Plan of Defence, drawn up in August 1939 a few weeks before the outbreak of war.

While British Foreign Office and Air Ministry memos frequently referred to the Italian embassy in Baghdad as "a nest of spies", the British press launched a campaign against the Iraqis' Italian-built aeroplanes. Questions were asked about why they were there in the first place, since a clause in the Anglo-Iraq Treaty stipulated that priority should be given to British aeroplanes if they could be delivered on time. An Italian offer to sell 15 IMAM Ro. 37s Lince reconnaissance biplanes was inevitably opposed by the British and on 12 April 1939 the British managed to have those RIrqAF officers who advocated these machines removed from the committee dealing with such matters. They were able to do this because the Iraqi Minister of Defence had announced that Italian equipment should be boycotted because of Mussolini's occupation of Albania a few days before, a position strongly supported by Wing Commander Cullay, the RAF's chief advisor to the RIrqAF who was another member of the committee.

In response, the Italian government announced that it would no longer supply spare parts, bombs, ammunition and other material needed to keep Iraq's SM 79Bs and Breda Ba 65 operational, although an order worth 3,241,000 lire was already in the pipeline. Whether any of these spares were acquired before the RIrqAF found itself in combat against the RAF as an ally of Italy and Germany in 1941 is unclear but seems unlikely. Six days after Mussolini and Hitler signed their "Pact of Steel" on 22 May 1939, official contacts between the RIrqAF and the Italian aero-industry came to an end.

Ten. Guza of the Italian Military Mission explaining the controls of a RIrqAF Breda Ba 65 (number 109) to the Iraqi Minister of Defence. The unusual shape of the front of the cockpit was designed to give the best possible view of ground targets. (Albert Grandolini collection)

This crisis came at a time when Iraq had already been thrown into a state of political uncertainty by the death of 26-year-old King Ghazi in a road accident on 4 April 1939. His four-year-old son became King Faisal II, with Ghazi's cousin and brother-in-law Prince Abd al-Ilah as regent. Iraq's monarch had earlier been suspected of having an extramarital, homosexual affair with an underage black servant – who was then "accidentally" shot in 1938 – the King's death merely adding to a sense of intrigue and scandal which already surrounded Ghazi. In fact, many in Iraq assumed that the British had engineered the Iraqi ruler's death, resulting in riots where the British consulate in Mosul was attacked and the British Consul murdered. Others suspected that Prime Minister Nuri al-Sa'id, in office since 25 December 1938, had "ordered" the accident because of a plan to unify Iraq and Kuwait. As an incidental consequence of this tragedy, the Iraqi Royal Flight's Percival Q.6, Percival Proctor and Miles Hawk were attached to No. 2 Squadron RIrqAF.

Sadly No. 3 Squadron suffered another fatality in April 1939, this time in England where *Mulazim Tayyar* Nuri Abu Tabikh had

Douglas DB-8A-4 of Royal Iraqi Air Force with a member of Iraqi royal family. This machine was probably serial number 140. (Albert Grandolini collection)

been sent for further training. While flying a Hawker Hart of the RAF he collided with an airborne target and crashed into the sea, his body never being found.

In the atmosphere of intrigue and simmering violence which followed the death of King Ghazi, the repair of the RIrqAF's first Savoia-Marchetti was completed in Italy in July. The Iraqi government agreed to pay a lump sum of £2,500 UK sterling for the work, plus £532 for the flight back to Iraq, and £360 for additional work uprating the Fiat A.80 engines. Next came the question of obtaining permission to overfly British controlled territories in the Middle East which proved more difficult than it had been for previous flights. Nevertheless, the aeroplane was eventually handed over to an Iraqi representative in Italy in August. Within days the Second World War erupted in Europe. Italy was initially neutral – and would remain so for ten months – so flight insurance had been difficult to arrange. Nevertheless, on 19 September *Tenente* Rodolfo Guza, who had previously trained Iraqi pilots on the Breda Ba 65, took off from Cameri aerodrome in the repaired SM.79B with co-pilot Algarotti, technician Maffezzoni and radio operator Sacco. They flew via Catania, Tripoli, Benghazi, Tubruq, Cairo and Lydda to reach Baghdad on 28 September 1939, Guza remaining as an instructor though the other crew members soon flew home in a civil airliner.

Just before war broke out in Europe, another Iraqi delegation travelled to the USA, looking at the possibility of purchasing American military equipment. It was headed by Nasir al-Janabi, CO of No. 1 Squadron who is understood to have been in command of the RIrqAF's Northern Command based at Mosul. The other members were Captain Midhat Abd al-Rahman (perhaps the same as Abd al-Rahman, Majd al-Din) and Captain Muhammad Hassan. They left for the United States in March 1939 and visited a number of American aircraft manufacturers, including the newly established Northrop Corporation at Hawthorne in California, mainly, it seems, looking for a dive-bomber – a form of attack currently regarded as the most accurate yet devised. As a consequence, al-Janabi and his colleagues recommended that the RIrqAF order 15 Northrop 8A-4Ds (a variant of the Northrop A-17) which, following the Douglas Corporation's take-over of Northrop, came to be known as the Douglas DB-8A-4 or Douglas Type 8 bomber.

On their way back from the USA, the two Iraqi officers reportedly stopped off in Italy and there expressed the RIrqAF's interest in buying five of the better-known three-engine versions of the Savoia-Marchetti SM 79 Sparviero. When informed of this request, the senior commander of the Regia Aeronautica gave his assent but insisted on prompt payment in cash. Nothing, of course, came of the proposal because the British made their opposition abundantly clear.

5

EGYPT, FROM EAAF TO REAF

With a new Anglo-Egyptian Treaty being negotiated, there were a number of important changes in Egypt's armed forces. These mostly impacted the Army rather than the Egyptian Army Air Force (EAAF), but the Air Force's senior men had already been promoted in 1935: Victor Hubert Tait to the rank of *Amir Alay* (*Miralai* or Colonel) and Jack Cottle to that of *Qa'im Maqam* (Lieutenant Colonel). Probably at this time junior officers were similarly promoted. April 1936 saw the departure of Air Commander Sir Christopher Joseph Quintin Brand, the Director General of Aviation in Egypt, whose secondment came to an end. In fact, Brand transferred to No. 6 Auxiliary Group RAF but before his departure the Egyptian government awarded him the Order of Isma'il on 30 March 1936. Air Commander Brand would subsequently play a major role in the Battle of Britain as commander of No. 10 (Fighter) Group responsible for the defence of south-west England and South Wales.

Just under one month later, *Qa'im Maqam* Jack Cottle MBE, DFC who now held the British rank of Squadron Leader, transferred back to the RAF where he briefly commanded No. 38 Squadron before rising to the rank of Wing Commander. His

Ibrahim Hassan Gazerine (centre) with two of his colleagues on the beach at Marsa Matruh. This was used as an airstrip when a flight of REAF Avro 626s was sent to patrol the frontier with Italian-ruled Libya in 1938 during the tension caused by the Munich Crisis. (Gazerine archive)

Pilot Officer Ibrahim Hassan Gazerine of the REAF with his observer-gunner and their Avro 626 on the beach at Marsa Matruh in the summer of 1938. A Flight of Egyptian Avro 626s were based here to patrol the frontier with Italian-ruled Libya during the Munich Crisis. (Ibrahim Gazerine collection)

place in the EAAF was taken by Squadron Leader N.P. Dixon AFC, a New Zealander who had served in Egypt both with the RAF Air Staff and at Heliopolis aerodrome. Victor Hubert Tait recalled some of the problems of this period of transition in Egypt, which included sorting out how Tait himself would hand over command. The British Inspector General of the Egyptian Army, Charles Watson Spinks, told Tait that he would have to leave immediately, as had other British officers attached to the Egyptian Army, but Tait replied that he only took orders from the Egyptian Minister of War. When Tait went to see the Minister, they discussed this problem and eventually agreed on what became the system of Air Advisors which came into effect in the spring of 1937.

In the meantime, Egypt continued to impose economic sanctions on Italy because of its ongoing invasion of Abyssinia. Even so, the international status of the Suez Canal meant that Egypt could not prevent the passage of Italian ships. Meanwhile the EAAF was at last able to make a small contribution to the defence of Egypt's frontiers. When interviewed by the author, Ibrahim Hassan Gazerine recalled how a Flight of Egyptian Avro 626s were sent to support the RAF's No. 208 Squadron, some of whose Hawker Audaxes were already patrolling Egypt's border with Italian-ruled Libya. While No. 208 Squadron operated from Marsa Matruh and Siwa Oasis, Gazerine was amongst the recently qualified Egyptian aircrew who were sent to Marsa Matruh.

Three Avro 626s normally patrolled together, though technical problems sometimes meant there were only two. The observer's position in one machine was occupied by a radio operator, while the others had Lewis guns mounted on their observers' cockpits, the first time that any Egyptian aeroplanes had been armed with guns. They would patrol at an altitude of 125m along the coast from Marsa Matruh to the frontier, the actual frontier from the coast to Siwa remaining the RAF's responsibility. Neither the British nor the Egyptian airmen met any trouble during their patrols.

Most of the REAF had to rely on old-fashioned methods of collecting and carrying urgent military messages, as demonstrated in this retouched photograph of an Egyptian Avro 626 pilot practising with a message retrieval hook around 1936. (EAF Museum collection)

The intensity of REAF training during the immediate pre-war year inevitably led to accidents. Here a damaged Avro 626 (number J327) has its wings and much of its fuselage fabric removed while being retrieved from typically stony desert terrain. (R.A. Hammersely photograph)

In Egypt the crisis caused by the Abyssinian War meant that a Second Egyptian Air Rally (in fact the third, if the one held at Heliopolis in 1910 is counted), planned to have been held in March 1936, was postponed until the following year. It would have included stops at the country's major archaeological sites and along the Red Sea coast. Instead, March 1936 saw four Egyptian civilian aeroplanes from the Misr Aviation School at Almaza make a formation flight to the Hijaz in Saudi Arabia, led by Sabri al-Kashif Effendi and Muhammad Sidqi. The latter had been involved in pioneering this route in December 1934, along with the idea of a scheduled air service to link modern technology with the annual Islamic *Hajj* or Pilgrimage (see Volume 4). This project had also hoped to establish a hotel and a bank in the Hijaz, presumably for the *Hajjis* or pilgrims.

According to Tarek Sidki, his grandfather's logbook shows that he first flew to Saudi Arabia in 1936. There he met King Abd al-Aziz Ibn Sa'ud for further discussions about the air service for pilgrims, and at the end of this visit the Saudi ruler gave Sidqi (note the slight difference in the transliteration of names between the different generations, as is often the case) a watch which is still in the family's collection. Sidqi was probably thus the first Egyptian civilian pilot to fly to the Hijaz. Misr Airwork of Egypt, or Misr Airlines as it came to be called, eventually signed a contract with the Saudi Arabian authorities to start the air service in 1937. Sadly, it was disrupted by engine problems while the costs were so high that the numbers of *Hajjis* willing or able to pay the fare were so few that the project was abandoned.

It was probably after this failure that Muhammad Sidqi found himself helping facilitate the continued manufacture of Klemm aeroplanes under licence in the UK. The machine in question was the B.K. Swallow, built by the British Klemm Aeroplane Company, the first of which flew in November 1933. Because the German-built Klemm L.25 proved popular in the UK, Klemm's salesman in Britain, Major E.F. Stephen, set up this British Klemm Aeroplane Company at the London Air Park at Hanworth in west London and built 28 B.K. Swallows. In 1935 this company changed its name to the British Aircraft Manufacturing Co. Ltd and went on to build 107 Swallow IIs (called the B.A. Swallow II reflecting the company's change of name), plus 36 more powerful B.A. Eagle IIs with enclosed cockpits.

It is worth noting that early in 1939 Saudi Arabia reached an agreement with the Misr Air Flying School for Saudi pilots

The REAF's second batch of Avro 626s had tail wheels rather than the earlier tail skids. Subsequently Egypt's surviving early Avro 626s appear to have been retrofitted with tail wheels. (V.H. Tait collection)

to be trained at Almaza on the outskirts of Cairo (see Chapter Three). Quite what Saudi students made of the very upper-class European atmosphere in the Misr Airwork Club House is unknown. It had been consciously modelled on what has been described as the "social life" found in the Air Works headquarters at Heston aerodrome in west London, both Heston and Almaza attracting the fashionable 1930s equivalent of what would later be known as the jet-set.

In April 1936 the German Ambassador to Egypt reportedly "got lost in the desert". In fact, Baron Eberhard Von Stohrer had been competing in a motor race from Cairo to Bahariya oasis when, on 18 April, he was caught in a dust laden *khamsin* wind from the Sahara Desert. Both the EAAF and the Egyptian Frontier Force were sent to find him. Sir Miles Lampson, the British Ambassador in Cairo, even asked the RAF to help and it took some days for Von Stohrer to be located before being rescued by Patrick Clayton, a British surveyor who worked for the Egyptian Desert Survey. Several years later, during the Second World War, some people suggested that the Baron intentionally "got lost" so that he could look for potential landing

The interior of a Westland Wessex showing what was, for its time, an advanced Marconi radio instillation. (Fleetway archive)

Ahmad Isma'il as a civilian flying cadet in Germany during the 1930s. After qualifying as a pilot, Ahmad Isma'il worked for Egypt's national airline, Misr Air but, like others who had learned to fly in Germany, came under suspicion by the British who unjustly feared he had pro-Axis sympathies. (Ahmad Isma'il collection)

grounds in Egypt's Western Desert, though this is almost certainly untrue.

Meanwhile the German government was eager to establish an air route from Berlin to the ex-German colonies in Africa, via Athens and Cairo. Under pressure from the British, but also out of caution about Nazi Germany's longer-term aims, the Egyptian authorities refused to grant the necessary landing facilities. The British Director of Egyptian Civil Aviation, Group Captain Bone, was particularly keen to avoid what was interpreted as German penetration of British-ruled airspace. According to Saul Kelly, in his book *The Hunt for Zerzura. The Lost Oasis and the Desert War* (London, 2002), Bone was opposed by Tahir Pasha, the Vice-Charman of Misr Airwork whom the British regarded as "pro-German". In reality, Tahir Pasha was entirely "pro-Egyptian" in wanting to minimise or dilute the British domination of his country.

Instead, the Germans suggested the formation of a joint Egyptian–German airline to be called *Horus*, using German financial backing and some German "experimental flights", along with an offer to sell German aeroplanes at much-reduced prices. The idea appealed to Tahir Pasha but not, of course, to the British and so too failed to get off the ground. In fact, Muhammad Tahir Pasha was said by the British Embassy, which obviously disliked him, to be not only "very much above himself" and "rather a dangerous person", but also to be acting as the King's spy within the upper echelons of Egyptian society. Speaking fluent English, French and German, he had been educated in Germany and had served in the Ottoman Turkish Army during the First World War.

Meanwhile, according to Saul Kelly, the Hungarian explorer László Almásy was Tahir's private pilot, worked as a flying instructor at Almaza, and encouraged Egypt's Royal Aero Club to acquire Hungarian-built aircraft and gliders. These might eventually have included two MSrE M-24 single-engine, single-seater "touring" aeroplanes purchased by King Faruq. To encourage such purchases, Almásy flew a glider around the Pyramid of Cheops, otherwise known as the Great Pyramid of Giza, located on the western edge of Cairo.

Ahmad Isma'il, with a white flying helmet in his hand, standing next to Egypt's best-known female pilot Lutfia al-Nadi. They are in front of one of the Misr Flying School's De Havilland DH 60 Moths at Almaza aerodrome in the late 1930s. Second from the left is Prince Abbas Halim, one of the flying school's patrons. (Ahmad Isma'il collection)

The Egyptians only had one Westland Wessex tri-motor transport (number W202) and for several years this rugged machine was called upon to carry out all the REAF's longer distance transport duties. (Author's collection)

On 28 April 1936 King Fu'ad of Egypt had died at the age of 68, being succeeded by his 16-year-old son Faruq. But, because of the young ruler's age, there was a Regency until King Faruq's coronation of 29 July 1937. Faruq came to be a controversial ruler but at the time of his accession there was widespread optimism in both Egypt and Britain that the Land of the Nile was entering a new and more hopeful age. Faruq was not only young but was thought to be a pious Muslim, which made him popular with the ordinary people. Indeed, he was known – for a while – as Faruq the Pious.

Although Faruq also presented himself as a "real Egyptian", in contrast to his "Turkish" forefathers, there was little basis to such a claim. His mother, Queen Nazli Sabri, had been of largely Turkish origin while also being descended from the famous French officer Joseph Anthelme Sève who, after fighting in Napoleon's navy and army, served Muhammad Ali, the founder of the Egyptian ruling dynasty. In Egypt, Sève converted to Islam and adopted the name of Sulayman Pasha al-Faransawi ("the Frenchman"). During

The first six Hawker Egyptian Audaxes to be delivered to the REAF, photographed at Almaza shortly after their allocation to No. 4 Squadron (Author's collection)

The first Hawker 674 or "Egyptian" Audax (number K400) to be delivered to the REAF was probably the original prototype upgraded with a Fairey Reed propeller, low pressure tyres and a tail wheel. (Ibrahim Gazerine collection)

the later years of his reign, especially during the Palestine War of 1948–1949, King Faruq sought to highlight his Arab identity, being described in the more loyal parts of the Egyptian media as "Faruq, beloved of the Arabs".

There had been many, sometimes conflicting, influences upon Faruq during his adolescence. One such was General Abd al-Aziz al-Masri, Director of the Police Academy, who had been one of the prince's tutors. In 1936 al-Masri was chosen by the Regency Council to be one of the group who would accompany King Faruq on a visit to England, though the young monarch quickly came to resent the General's strict discipline. Another elder and hopefully steadying influence upon the increasingly extravagant and wilful young King was Muhammad Ahmad Hasanain. A more cultured and diplomatic man than Aziz al-Masri, Hasanain would remain close to King Faruq.

Even before the death of King Fu'ad there was a subscription campaign to raise money for new aircraft for the EAAF. Begun by Egyptian students with the aim of purchasing three squadrons of aeroplanes, it proved a considerable success. Members of the government donated a month's salary, inspiring or shaming members of the two houses of parliament and senior civil servants to do the same. Several very rich people offered substantial sums, while hundreds of thousands of ordinary people in town and country gave what they could.

It would take time for this public effort to produce results but in the meantime the EAAF acquired an Avro 641 Commodore. It was a luxurious biplane with an enclosed cabin that could carry four passengers in addition to its pilot. Powered by an Armstrong Siddeley Lynx IVC radial engine, the Avro 641 was of all metal construction and had a cabin comparable that of a high-class motor car. It was also a rare machine, only six being built. Two were sold to Egypt in September 1935, where they were given civilian registrations SU-AAS (c/n R3/CN/700, ex-G-ACRK) and SU-AAU (c/n R3/CN/721, ex-G-AAU). The former, SU-AAS, had been owned by Victor Hubert Tait and was transferred to the Royal Egyptian Air Force after the Tour of the Oases Air Rally (see below) and was given the military serial number W203. It has been suggested that the second Egyptian Avro Commodore was also transferred to the REAF, but there is no hard evidence. While SU-AAS became W203, W202 was the Egyptian Air Force's Westland Wessex, W204 the Avro 252, while W205 to W207 were Anson Is delivered shortly before the outbreak of war. No higher numbers in the W range are known, though it is possible that a second Avro 641 (ex-SU-AAU) became W208. Indeed, the author has a hazy memory of seeing a model Avro 641 in the Military Museum in the Cairo Citadel in 1964, painted silver but with the serial number W208 which he subsequently assumed to be incorrect.

The substantial money raised by public subscription was used to buy a squadron of Avro 674s, which were licence-built versions of the Hawker Audax widely known as "Egyptian Audaxes". Their purchase was negotiated in 1936 and deliveries began in 1937. An increasing number of pilots would also be needed to operate Egypt's expanding Air Force, and one of those who completed their training in 1936 was Abd al-Hamid Sulayman. He later became commander of Hilwan aerodrome, home of Egypt's fighter squadrons prior to the Palestine War.

De Havilland DH 83 Fox Moth (SU-ABG) of Misr Airwork
This Fox Moth (c/n 4024), powered by a De Havilland Gipsy III inverted inline engine, was sold to Misr Air in August 1935 and was given the Egyptian registration SU-ABG. It was subsequently sold to a British owner and was given a new registration, G-ADNF. Later still the machine was sold to Australia where it received yet another registration, VH-ABQ, and was eventually "lost to enemy action" at Rabaul (now in Papua New Guinea) in January 1942. There were several versions of the De Havilland Fox Moth, this example having a sliding canopy over the cockpit plus a substantial fairing along the top of the fuselage. Painted a typical Egyptian overall white with green markings, there was also the Misr Airwork company name and information written in English and Arabic on the front fuselage. However, unlike some later Misr Airwork and Misr Airlines aircraft, there is no writing on the tail. (Artwork by Peter Penev)

Avro Tutor (K3337) RAF No. 4 FTS Abu Suwayr
This aeroplane has minimal markings and there is no Avro manufacturer's logo on the side of the front fuselage. It is otherwise painted an overall silver dope except for a bare metal cowling and panels behind engine. It has normal British national markings. Like so many of the aeroplanes at No. 4 FTS in the 1930s, K3337 was flown by Royal Egyptian Air Force students as well as aspiring young pilots from the British RAF. Note that there is a grey rubberised canvas "blind flying" hood over the rear cockpit; this being shown in an open position. Also note the broad "balloon tyres" for use in the Middle East. (Artwork by Tom Cooper)

Meindl/van Nes A.VII Ethiopia, 1936
Designed by Ob.-Ing. Erich Meindl of the Burgfalke Flugzeugbau, this aeroplane was a modified version of the basic Meindl/van Ness A.VII Cadet or Meindl M7. It was delivered in kit form to the Ethiopian government, where it was given the name Tsehai (meaning Sun in Amharic), having been named after the Emperor Haile Selassie's third daughter Princess Tsehai. It was the first aeroplane to be assembled in Ethiopia. Some modifications were made during assembly because Addis Ababa was at a high altitude, including the addition of flaps along the entire wings and it was also given dual controls for use as a trainer. The undercarriage and propeller were imported from Germany. Painted overall silver dope, the machine had Ethiopian Air Force markings on the rudder and wingtips while the name Tsehai was written on the fuselage in Amharic lettering. The type-name and number ETHIOPIA 1 was written on the yellow stripe of the rudder insignia in both European and Amharic script, this having been intended as the first of three machines, though the others were never built. The solitary Meindl/Van Nes A.VII Ethiopia was taken to Italy following the Italian conquest and was exhibited in the Caserta Aviation Museum until 1941 and it was then taken to the Italian Aviation Museum at Vigna Valle. (Artwork by Tom Cooper)

Junkers 52/3m, Spanish Nationalist Air Force, 1936
One of twenty Junkers Ju 52/3m transports which the Nazi government of Germany offered to the Nationalist rebels at the start of the Spanish Civil War in 1936, this aeroplane arrived with air and support crews drawn from Lufthansa and the Luftwaffe. Furthermore, the aeroplane retained a German civil registration on the tops of its wings for a short time. The Nationalist Ju 52/3m transports promptly played a major role carrying pro-Franco troops, including many Moroccans, from the Spanish Protectorate in northern Morocco to southern Spain. They were almost certainly unarmed at this stage. This machine (serial number 22 * 79) is overall pale grey-green (probably Luftwaffe Grau 63), plus a black anti-dazzle area behind the central engine. The wing-engine cowlings are also black, but these early machines did not have the engine fairings on the wings painted black, as would be seen later. Nor did they have black stripes across the upper and lower surfaces of the wings behind the wing engines. A short while later the aeroplane would be roughly painted a darker tone, though still with the black anti-dazzle area, and a serial number would be repeated in small white numerals on the tail-fin. The aeroplane shown here has the earliest form of Nationalist identity markings, consisting of original Spanish red-yellow-red roundels plus a black saltire cross on a white ground on the rudder. (Artwork by Peter Penev)

De Havilland DH 82A Tiger Moth of RIrAF (71) c.1936
Painted an overall silver dope, this Iraqi Tiger Moth's serial number 71 has been added in black Arabic numerals on the fuselage behind the distinctive RIrqAF triangular national insignia, but with a European number repeated much smaller beneath the Arabic number. The usual Iraqi national markings consist of four vertical stripes on the rudder, plus the Iraqi triangle on fuselage sides, as well as below the lower and on top of the upper wings. This aeroplane has the normal short exhaust stub below the engine cowling seen on almost all military Tiger Moths, as well as normal tyres; there being no evidence of the use of desert "balloon tyres". (Artwork by Peter Penev)

Avro 641 Commodore (SU-AAS) of V.H. Tait, with Circuit of the Oases Air Rally number 41, 1937
This aeroplane (c/n R3/CN/700) was first registered in the UK as G-ACRX in May 1934, when it was owned by the Fifth Earl of Amherst, Viscount Holmesdale, and was based at Heston west of London. Viscount Holmesdale was an airline director as well as a pilot. The Avro Commodore was then transferred to Airwork, also at Heston, before being sold to the Canadian airman, senior RAF officer and first Director of the Egyptian Army Air Force, Victor Hubert Tait. It was then given the Egyptian civil registration SU-AAS. Only six examples of the Avro 641 Commodore were built, and were originally powered by Armstrong Siddeley Lynx IVc engines. Described as notably "luxurious", this machine would retain its over silver aluminium and polished metal finish for several years, though its markings would change when it was transferred to the REAF. The machine is shown here with the Circuit of the Oases Air Rally number 41 on its now white-painted rudder. During the rally it was flown by Ahmad Nagi. (Artwork by Peter Penev)

Waco UIC (SU-AAN) of Prince Abbas Halim
Waco UIC c/n 3764 was sold to Prince Abbas Halim at an unknown date and was flown by the prince during the Tour of Oases Rally in Egypt 1937. During that event it had the rally number 22 in black numerals on a white panel, either painted or pasted on the tail. Abbas Halim was one of two members of the extended Egyptian royal family to take part in this Tour of Oases Rally, the other being Prince Umar Halim flying a Miles M.2F Hawk Major (SU-AAP). This version of the Waco Standard Cabin Series had a V-shaped rear cabin window or skylight behind the upper wing. It also has multiple cylinder-head bulged fairings around the cowling of its powerful Continental R-670 engine. The colours of Prince Abbas Halim's Waco UIC are unknown, but most of these American machines were delivered in a pale cream or off-white colour, and given the Prince's vigorously patriotic sentiments, the trim and probably also the registration are most likely to have been green. (Artwork by Peter Penev)

Westland Wessex
Westland IV Wessex c/n. WA2152A was the last of this type of transport aeroplane to be built. It was powered by three Armstrong Siddeley Genet Major 1A (Civet I) engines and replaced one of the EAAF's Avro 618s which had crashed. This machine previously had the British civilian registration G-ACIJ; it was then given the EAAF (and subsequently REAF) serial number W202. The small writing on the tail and rudder, as well as a dark outline around the tail-plane, rudder and fuselage seen on the aircraft in its previous civilian guise, were removed when it was transferred to the Egyptian military and it was then painted an overall silver aluminium dope and had standard Egyptian national markings. There was, however, a small gold crown in the outer green ring of the fuselage roundel, though not on the wing roundels. In Egyptian military service the Wessex was fitted with balloon tyres so that it could operate from virtually uncleared desert airstrips. It was, in fact, said "to land like a cushion". (Artwork by Peter Penev)

Douglas DC-2B 115-F of LOT Polish Airlines (SP-ASL) on the Poland-Palestine route, 1937
Three DC-2Bs were sold to LOT Polish Airlines and were fitted with 750 hp Bristol Pegasus VI (560 kw) radial piston engines. However, one of them (registration SP-ASJ) crashed in Bulgaria in November 1937. They were overall polished bare metal with the dark blue LOT insignia on the tail. The Polish civil registration SP-ASL is believed to have been in black lettering on the rear fuselage and on the wings. THE POLSKIE LINJE LOTNICZE „LOT" lettering on the side of the fuselage, and over the fuselage door as „LOT", were probably also black. The first regular passenger service to use the new Palestinian airport at Lydda had been the Egyptian Misr Airwork service between Cairo and Nicosia via Lydda in August 1935. However, LOT used its DC-2Bs to open up a long-distance route from Warsaw to Lydda via Lvov, Czerniowice, Bucharest, Sofia, Thessaloniki, Athens and Rhodes; this being inaugurated on 4 April 1937. It was regarded as one of longest routes "in Europe" and linked the Zionist settler community in Palestine with the substantial Jewish community in Poland. (Artwork by Goran Sudar)

De Havilland DH 89A Dragon Rapide, Iraqi Aviation Association
This De Havilland DH 89A Dragon Rapide (c/n 6415) had received its certificate of airworthiness on 25th July 1938, and on the same day it gained its Iraqi civil registration YI-HDA. It was one of three Dragon Rapides registered to the Baghdad Aeroplane Society on the same date. A peculiarity of Iraqi civil registrations was that the alphabetical sequence is not the same as the chronological sequence. The writing on the nose, as it appears in the only available blurred photograph has been translated as Airlines of Iraqi Aviation Association, which was either the correct name of the Baghdad Aeroplane Society or was a later name of the same organisation. The Baghdad Aeroplane Society's three Dragon Rapides were reportedly transferred to the Royal Iraqi Air Force during the Second World War Two, but no military serial numbers are known. The machines were overall aluminium silver dope and bare metal. The civilian registration was written in black on the rear fuselage, being repeated beneath and above the wings. The writing on nose was probably also black. (Artwork by Peter Penev)

De Havilland DH 60 Gipsy Moth of Peake Pasha in Transjordan (G-ABMX) 1937
This De Havilland DH 60 Gipsy Moth was first registered as G-ABMX to an unknown owner in May 1931. It was subsequently sold at an unknown date to F.G. Peake (Peake Pasha), the senior officer of what became Transjordan's Arab Legion. The machine was then based in Amman, Transjordan, before being sold to Moshe Katz of Palestine Flying Services based at Lydda in Palestine in 1937. Throughout this period the aeroplane retained its British serial number and the same decoration until it was eventually written off charge. Thereafter, the fuselage, minus its presumably damaged wings, appeared with the same British registration and the same decorative colour-scheme when it was used for ground instruction in the newly declared state of Israel in 1948. The aeroplane had the standard De Havilland DH 60 Gipsy Moth decorative pattern in which an owner chose the two primary colours, one of which was usually silver dope. In this case the second colour is understood to have been red. Quite when the red and white chequer-board decoration on the fin was added is unknown, but as it seems to have reflected the red and white kifaya or head-cloth worn by the Arab Legion (and indeed most Transjordanians), it is assumed to have been when the aeroplane was owned by Peake Pasha. (Artwork by Peter Penev)

Helmy Aerogypt Mk. I (G-AFFG), 1939
Only one example of the Helmy Aerogypt "Safety Plane", designed by Egyptian engineer Mr. Salih Hilmi, was built. This same airframe was subsequently modified as the Aerogypt Mk. II, Mk. III and Mk. IV. It first flew at Heston aerodrome in west London in 1938 and originally had a hinged cabin roof which, when raised, acted as a landing flap. The version shown here is the Aerogypt Mk. I and is shown with the additional aerofoil-landing flap above the fuselage in the closed position. The machine was powered by three 22 hp (16kw) Douglas Sprite horizontal two-cylinder air-cooled engines based upon the Douglas F/G31 motorcycle engine. Its colour-scheme is unknown but the machine was either overall white or overall silver dope, with a darker coloured trim. The latter might be red, given how dark it looks in the photographs, but bearing in mind how patriotic the educated elite of Egypt was at this time, a dark green has been chosen for this reconstruction. (Artwork by Peter Penev)

Miles M.14 Magister (YI-GFH) of King Ghazi of Iraq, c.1939
This aeroplane (c/n 797) is a Miles M.14A Magister which had a taller tail than the original version, as well as anti-spin strakes ahead of the tailplanes. It was purchased by King Ghazi of Iraq on 15th August 1938. It retained its original overall silver dope except for some minimal decorative elements and the name SHATT AL-ARAB painted on the sides of the engine cowling. The colour of the fuselage lightning stripe is unfortunately unknown as only monochrome photographs are available. But assuming that it was in one of the Iraqi national colours of black, red, green and white, as they appear on the Iraqi flag on the aircraft's tail, the stripe seems most likely to have been a deep green. This colour has also been assumed for the registration YI-GFH and the name SHATT AL-ARAB. Here the central part of the otherwise varnished wooden propeller is in the same green, along with the small spinner. The type's Hawk motif below the windscreen is light blue with a yellow beak, outlined and picked out in black on a silver background. The letters were usually yellow but on the silver dope of King Ghazi's aeroplane they are written on a dark, perhaps black, disc. (Artwork by Peter Penev)

Fokker F.XVIII (VQ-PAF) Commercial Aviation Corporation, Lydda, Palestine 1938
This aircraft is a Fokker F.XVIII (c/n 5310), powered by three 313 kw (420 hp) Pratt & Whitney Wasp C radial engines. It was originally registered as PH-AIQ in the Netherlands in 1932 and was operated by KLM (Royal Dutch Airlines). In October 1936, KLM's F.XVIIIs were withdrawn from the service from the Netherlands to Batavia (now Jakarta in Indonesia). PH-AIQ and PH-AIR were then sold to ČSA (Československé státní aerolinie) when they were re-registered as OK-AIQ and OK-AIR in November 1935. This F.XVIII was next registered to the Commercial Aviation Corporation based at Lydda aerodrome in Palestine, where is received the Palestinian mandate registration VQ-PAF. However, the company only seems to have had one aircraft, using it for a short while before it crashed at Lydda on 13 January 1939. Apparently, the Fokker was not destroyed as it remained on the airfield until 1948 when it was taken on charge by the Israeli Air Force. Nevertheless, it was never re-built and was eventually scrapped. In Palestine this Fokker F.XVIII seems to have retained the same blue and silver colour-scheme it had when owned by the ČSA. However, the Československé státní aerolinie writing on the sides of the front fuselage and nose was removed, as had that of KLM previously. A new Palestinian registration (VQ-PAF) was added in black lettering on white panels on the rear fuselage and under the wings. Small spinners had also been added to the propellers. To further confuse matters, the registration VQ-PAF is said to have also been given to a RWD-13 which previously had the Polish registration SP-BFR and was named Yemen 2. This was registered to the Palestine Flying Club and to the Aviron Palestine Aviation Co. (Artwork by Peter Penev)

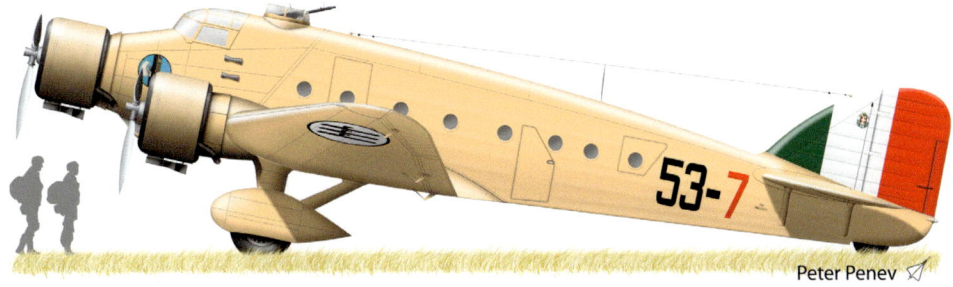

Savoia Marchetti SM 81, Italian Air Force, 53rd Squadriglia of the 15th Stormo, Castel Benito, Libya 1938
The three-engine, fixed undercarriage Savoia-Marchetti SM 81 bomber was a militarised version of the SM 73 airliner. It was later selected as Italy's first paratroop transport. It was in the machines of the Regia Aeronautica's 15th Stormo that the Ascari del Cielo (indigenous Libyan paratroopers) trained at Castel Benito aerodrome outside Tripoli. Here, on 16 April 1938, three-hundred Libyan volunteers made their first coordinated drop from twenty-four aeroplanes, watched by Marshal Italo Balbo, the Governor-General of Libya. A few days later another drop was made at night. The version of the SM 81 transport used by the 15th Stormo had Gnome Rhone K.14 engines with three-blade propellers. Air-intakes for oil filters under the engines appear to be longer than usual, perhaps because these aircraft were operating in North Africa. Here the aircraft is also shown with its upper gun turret in the raised position. The aircraft is overall bianco avorio 5 and has old-style green-white-red tricolori stripes in the rudder. Being part of the 53rd Squadriglia it bears the identification number 53-7, plus the type identification and MM serial number in small black lettering on the rear fuselage. The actual MM number is, however, unknown. (Artwork by Peter Penev)

Vickers Wellesley Type 292, RAF Long Range Development Unit (LRDU) 1938
A flight of three Vickers Type 292 Type 292 Wellesleys (L2638, L2639 and L2680) of the RAF's Long Range Development Unit (LRDU) achieved a non-stop, two-day flight from Ismailia in Egypt to Darwin in Australia in November 1938. L2638 and L2680 thus established a new distance record of 11,526 km but L2639 had to divert to refuel in the Netherlands East Indies. Shortly afterwards it also made an emergency landing and was damaged beyond repair. A fourth LRDU Wellesley (L2681) formed part of the original group of aircraft which flew from the UK to Egypt but this machine did not take part in the Australia flight. The Type Vickers 292 had been modified for long-distance flying by the LRDU. The alterations included a fuel dumping system allowing an emergency landing when the weight of fuel would otherwise have caused the aircraft to exceed its maximum landing weight. The main external difference was a longer engine cowling than that on ordinary Vickers Type 287 Wellesley Mk. Is. In most respects the appearance of the LRDU's Wellesleys was the same as those in squadron service, but the machines which made the record-breaking Egypt to Australia flight also had RAF badges on their noses. (Artwork by Goran Sudar)

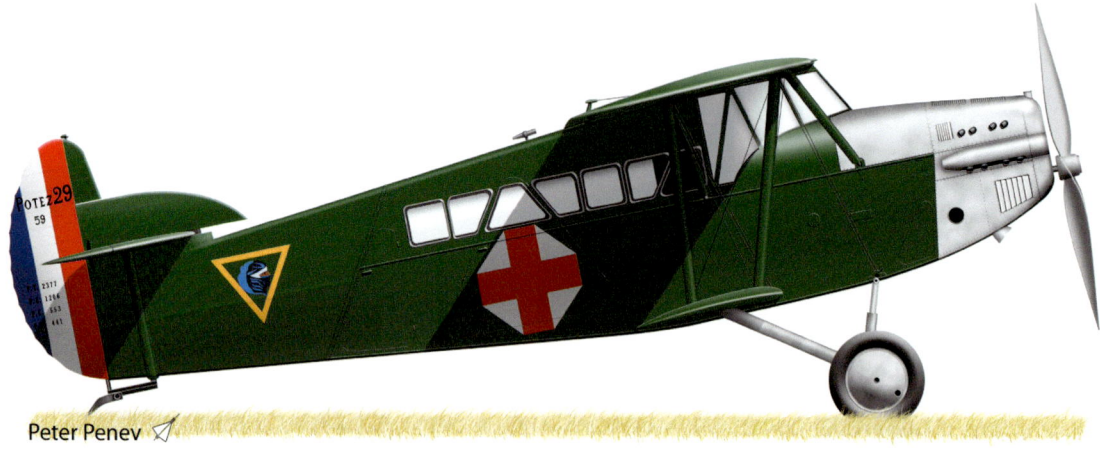

Potez 29.10 Sanitaire, Escadrille VR 548, French Algeria
This Potez 29 Sanitaire was part of Escadrille VR 548 of the 2nd GAA based at Wahran (Oran) before the Second World War. It was also used as a light transport when the squadrons of the 2nd GAA changed locations. It is painted an overall dark green vert protectif, including the inter-plane struts. The only exceptions appear to be the engine cowling, undercarriage struts and wheel hubs which are unpainted aluminium. The insignia of Escadrille VR 548 consisted of the head of a veiled Tuareg warrior in an inverted red triangle bordered with yellow or yellow-orange. The lettering on the rudder is in the usual Armée de l'Air style of this period and consists of the type indication followed by smaller technical information. (Artwork by Peter Penev)

De Havilland DH 82A Tiger Moth, Imperial Iranian Air Force 1938
This is one of the last batches of Tiger Moths to be delivered to Iran and retains its original overall silver aluminium dope paint scheme. The Iranian Imperial (Shahinshah) crown on the sides of the fuselage is gold with black details. The black serial number 152 on the rear fuselage, on the white stripe of the Iranian national markings on the rudder, on top of upper wings and underneath lower wings, is in Persian-style numerals which differ in some respects from the numbers used in Arab countries. Note that the numbers above and below the wings all read from the rear. (Artwork by Peter Penev)

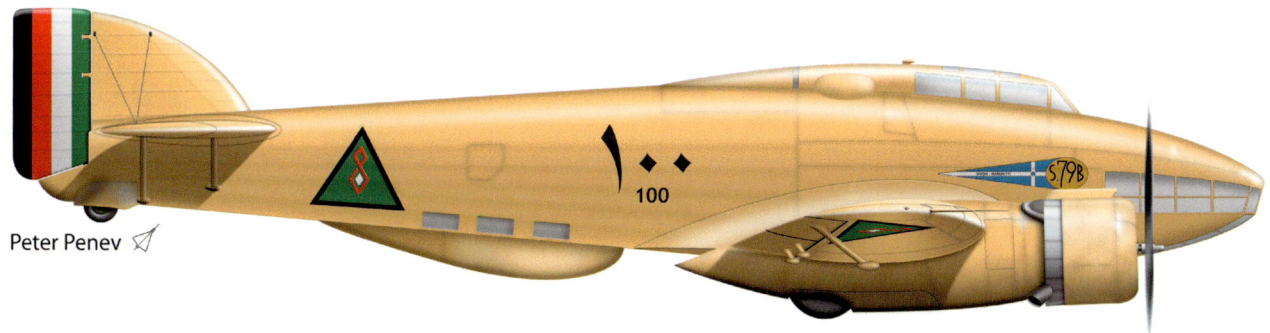

The first Savoia Marchetti SM 79B (number 100) delivered to the RIrqAF
The first Savoia Marchetti SM 79B to be delivered to the Royal Iraqi Air Force was serial number 100. It lacked the glazed panels on top of the aircraft's nose which were seen in subsequent aircraft. It was also the only Iraqi SM.79B to have a distinctive type badge with an extended pennon on the sides of the front fuselage. Like the later aircraft, it was painted an overall sandy yellow which was probably Regia Aeronautica giallo mimetico on both the upper and lower surfaces. The black serial number 100 is painted in large Arabic and smaller European numerals on the sides of the fuselage, and the national markings are the standard RIrqAF triangles, plus four vertical coloured stripes on the tail. (Artwork by Peter Penev)

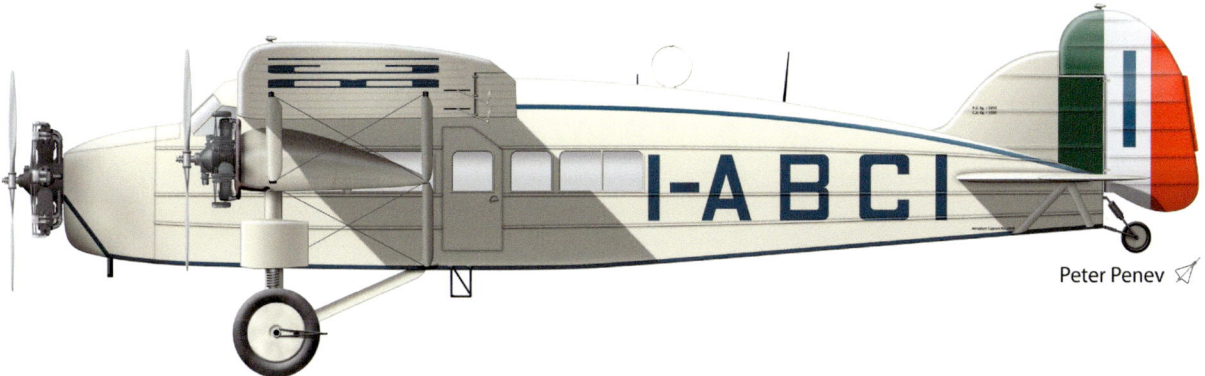

Caproni Ca.101bis of the Saudi Arabian AF (I-ABCI) 1937
The Caproni 101 bis aircraft sold to the Saudi Arabian government had previously been owned by the Italian airline, Ala Littoria. They clearly retained their Ala Littoria colour-scheme, the Italian civil registration (I-ABCI) and even the Fasces symbols for some time while in Saudi service. In fact, it is not known whether they were ever repainted. The Ala Littoria colour-scheme of this period was an overall off white or pale cream with dark blue trim, plus stripes along the four corners of the fuselage but not on the tail or wings. The Italian civil registration I-ABCI on the fuselage and beneath the wings is also dark blue while the large vertical stripe or capital letter I in the white stripe of the Italian tricolour on the rudder is probably also dark blue. The main visible between the Caproni Ca.101bis airliner and the military versions are the former's slightly larger dimensions, larger rudder, more angled but not entirely squared off wing-tips and significantly larger shock absorbers on the main undercarriage struts. The airliner also has larger, squared fuselage windows and different engines. The central engine of the Ca.101bis was an Alfa Romeo Jupiter, licence-built version of the Bristol Jupiter, while the left and right underwing engines were Armstrong Siddeley Lynxes. As an unmodified ex-civilian airliner this Caproni Ca.101 of course has no armament. (Artwork by Peter Penev)

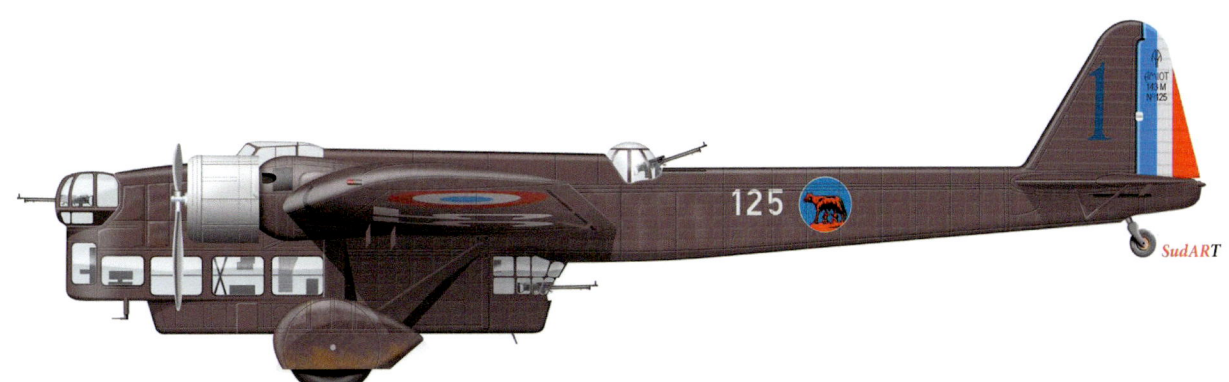

Amiot 143, French Armée de l'Air, 4th Escadrille of GB II/38, French Algeria 1939
This Amiot 143 (number 125) was part of the 4th Escadrille of GB II/38 based at Setif (Stif) in Algeria in 1939. Following an accident on 12th July 1939, it was sent back to France at the start of the Second World War and was then based at Troyes-Barberey. The aeroplane is in an overall dark brown chocolat except for the engine cowlings and the underneath of the engine nacelles. There are French national markings consisting of a tricolour on the rudder and roundels under the wings, but not on the fuselage. The fuselage roundel has, in fact, been replaced by the unit insignia which consists of an orange-red wolf nourishing Romulus and Remus, standing on orange-red ground, on a relatively dark blue disc with a narrow white edge. There is also a very large white number 55 under the wing, read from the rear, plus a large and relatively dark blue number 1 on the tail fin. The machine's individual number 125 is in white on the sides of the fuselage, and in black as part of the lettering on the rudder. The latter also includes the type identification and the motif of the Amiot company. (Artwork by Goran Sudar)

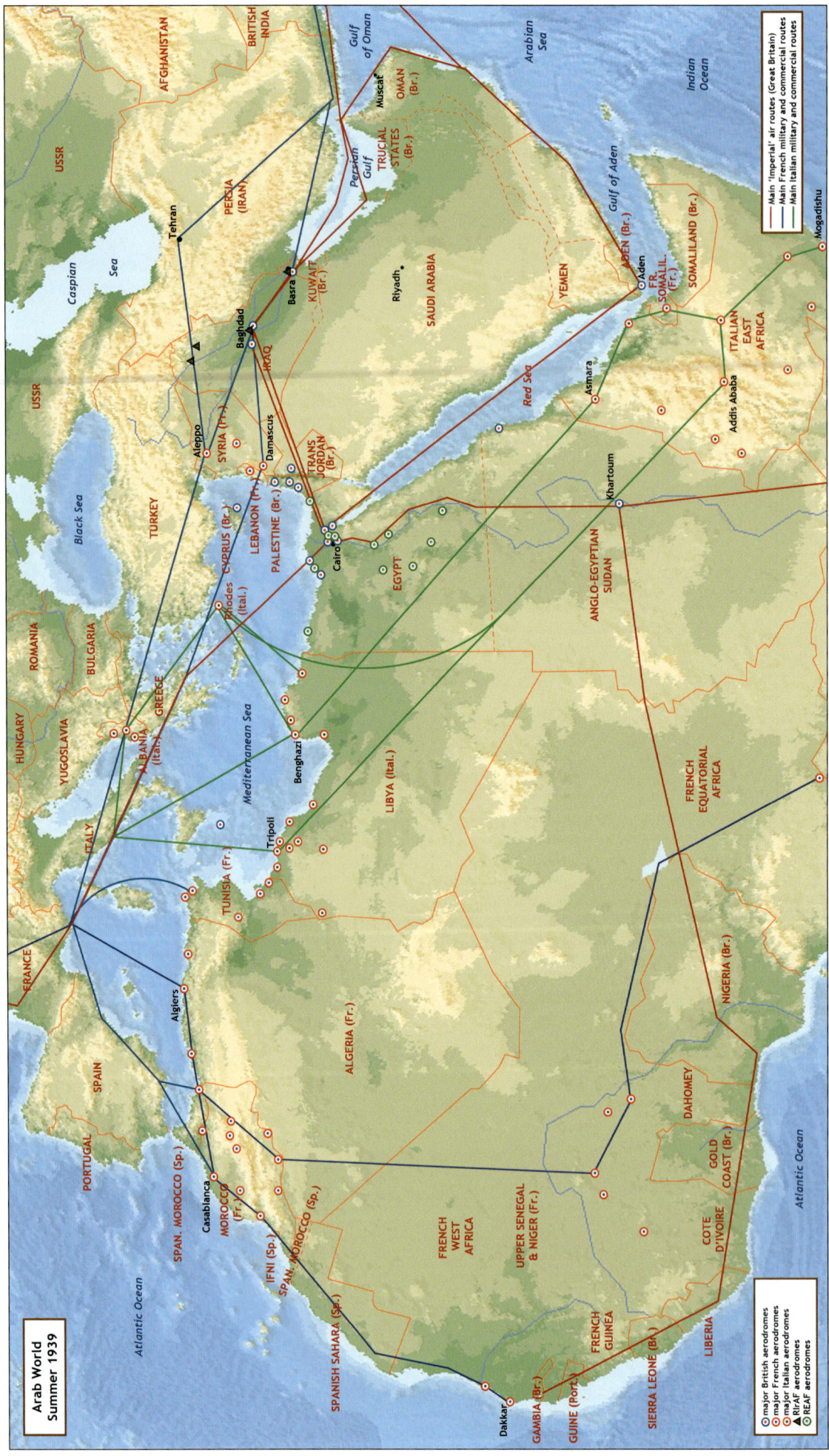
(Map by Tom Cooper)

By July 1936 it was clear that Italy was going to win the Abyssinian War, so the British decided that the international crisis should be "wound down". The RAF's No. 208 Squadron ceased its frontier patrols from Marsa Matruh and returned to its base at Heliopolis. This left the EAAF to patrol the Libyan frontier on its own, though these flights probably also ceased after a while. The situation in Palestine was meanwhile getting serious, with growing violence between the Arab inhabitants and the Zionist settlers. 208 Squadron was therefore told to prepare two Flights for a possibly hurried redeployment to Palestine.

Egypt's first order of Hawker "Egyptian" Audaxes included the upgraded prototype and number K401 which is seen here flying over the southern end of Gezira Island in Cairo. These machines were powered by Armstrong Siddeley Panther VIA engines. (Ibrahim Gazerine collection)

There were several reasons why a new Anglo-Egyptian Treaty was necessary. The most important was Britain's need for better relationships in the Arab Middle East in the wake of Fascist Italy's crushing of Sanussi resistance in Libya and its conquest of Ethiopia. On the Egyptian side there was increasing concern that Italy might try to seize a land corridor across northern Sudan to link Italian possessions in Libya with its expanding territory in East Africa. This was certainly a matter of concern to Mustafa Nahas of the Wafd Party who had won a general election and returned to power as Egypt's prime minister on 9 May 1936 (which he remained as until 29 December 1937). Even if such a menace was unrealistic – at least in current circumstances – there was a genuine possibility that Egypt and the Sudan could be enveloped by Mussolini's growing African Empire.

Pilots of the REAF's first Audax squadron in front of one of their new machines, probably in 1937. (EAF Museum collection)

One of the primary aims of the new treaty was to strengthen Egypt's armed forces so that they could defend the Suez Canal and cooperate effectively with British forces. With so little time until the outbreak of the Second World War, it is not surprising that such efforts achieved only partial success. Nevertheless, increasing recruitment resulted in a more diverse officer corps, though a substantial number of those enlisted during this period came to have a stronger loyalty to their country Egypt, than to its ruler King Faruq. They also came from a highly politicised generation of young men like Gamal Abd al-Nasser and Anwar Sadat who viewed Britain with profound mistrust. The latter was seen as an imperialist power which somehow continued to occupy and politically dominate Egypt, despite the country's legal independence.

Following the Wafd Party's return to power, Prime Minister Nahas Pasha headed a delegation, representing several parties in addition to his own, in negotiations with Great Britain about a new Anglo-Egyptian Treaty. Agreement was reached in August 1936 and the new treaty was signed on the 26th. It marked a significant step towards true independence, though Britain retained a number of military rights within Egypt by which the British intended to preserve the vital interests of their empire which was, of course, still a major power in the world. Some other changes were more symbolic than real, with the position of British High Commissioner now becoming British Ambassador. As such he still had significant powers of influence, persuasion and, when Britain thought necessary, bullying.

The continued presence of British forces on Egyptian soil was justified on the grounds of defending imperial communications

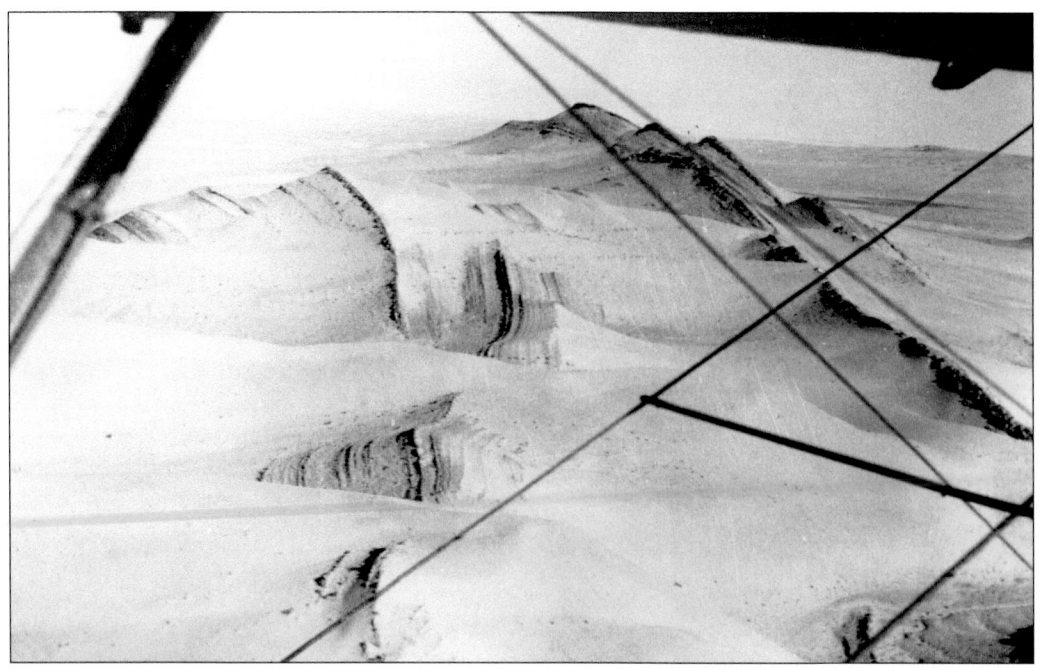
The mountains of western Sinai in 1938, not far from the Suez Canal and probably photographed from an Audax. (Woodroffe family archive)

On the political level, both the British authorities and the Wafd government claimed that the 1936 Anglo-Egyptian Treaty would result in an improvement in relations. There were, however, men on both sides who had doubts. Some Egyptians had little faith that the British would keep their word, amongst whom were the religiously based Muslim Brotherhood and the quasi-fascist Young Egypt Movement. Meanwhile on the British side there was a widespread assumption that Egyptians would always bow to British influence or persuasion. There were also those whose prejudice against the Egyptians ran so deep that they disapproved of the whole idea of handing over political or, worse still, military power. In practical terms the continuing presence of large numbers of British soldiers, airmen and to a lesser extent naval personnel in Egypt reinforced local mistrust, while ensuring that if a crisis did arise British military intervention could be swift and relatively easy.

– above all the Suez Canal. Indeed Article 8 of the new Anglo-Egyptian Treaty specifically allowed Britain the right to keep 10,000 troops and 400 aircrew in Egypt in time of peace. They would remain in the area of the Suez Canal for a further 20 years. The RAF still had the right to overfly Egyptian territory for training purposes and during communications flights. In case of war with a third party, Britain and Egypt would provide substantial "mutual assistance", which in practice meant British troops returning to Egypt, and it was agreed that the new treaty would be extended after 1956. Naturally no one foresaw that 1956 would see British-Egyptian relations collapse into conflict.

Prior to the signing of the new treaty there was no clearly defined "Canal Zone", but this now began to emerge around military airfields and their supporting infrastructure. Furthermore, under Article 8 of the Treaty, it was clear that although the Egyptian Army would eventually be responsible for securing freedom of navigation through the Suez Canal, this would only happen when Egypt was capable of doing so. The decision as to when this status was reached was down to the British – if not necessarily in such blunt terms. The British Inspector General of the Egyptian Army, a post which itself replaced the *Sirdar* back in 1924, now disappeared with an Egyptian senior officer being placed in charge of the Army under the Minister of War and the King. Nevertheless, the British remained in actual control of training and the selection of weaponry for Egypt's forces.

The future fate of the Anglo-Egyptian Sudan similarly remained unspecified. This proved to be a hostage to fortune, as there was already a yawning gulf between Egyptian and British views on the subject; the Egyptians clung to their traditional policy of "the unity of the Nile Valley". Nevertheless, the Wafd government and party were justified in presenting the treaty as a significant success for Egyptian negotiators and for their country as a whole. They also made clear that they would now strive for a speedy strengthening of all Egypt's armed forces which would, they claimed, lead to a speedy evacuation of British forces. This was undoubtedly believed by most of the next intakes of cadets to the Egyptian Military School.

Whether the return of No. 208 Squadron RAF from Marsa Matruh to its former home at Heliopolis outside Cairo around August 1936 was connected with negotiations for a new treaty is unclear. Perhaps the British wanted to emphasise Egypt's continuing dependence upon the RAF in the face of a potential Italian attack, though it is more likely that the British authorities already had their eyes upon the deteriorating situation in Palestine. A few weeks later in late September 1936, No. 208 Squadron was sent to Jerusalem and Haifa.

Whatever the reality, the new Anglo-Egyptian Treaty meant that the British military occupation of Egypt officially came to an end. The finer details of the treaty stated that British troops would normally be confined to Sinai and along the Canal. Here the Egyptians would build barracks according to British plans, but at their own expense. The British could also move anywhere and could conduct reconnaissance right up the edge of Giza Governorate – in other words up to and including Cairo but no further south. For many decades Britain had imported more from Egypt than it had exported, but now it was hoped that this trade imbalance would be removed by selling to the Egyptians substantial amounts of military equipment. There was, of course, an assumption that the Egyptian armed forces would "buy British" while continuing to rely upon British military instructors. The British were less happy when General Aziz al-Masri was chosen as the new, if largely nominal, head of the Egyptian Army, replacing the British Inspector General.

Another important symbolic change was that Egypt could send military attachés to Egyptian embassies around the world. Although the future of the Sudan remained undecided, the British agreed that Egyptian troops could return to there under their own officers. Thus, when an Egyptian military column left for the Sudan, the entire country celebrated. Similarly, the Egyptian intelligence services, which had been run by British officials, were handed over entirely to Egyptians, which was more significant than most people realised.

One immediate impact of the new treaty was upon military recruitment, with Egypt's expanding armed services needing many more officers. Partly as a result of the policies of the Wafd government, and partly because of simple practical considerations, Egyptian officer cadets now started to be drawn from a broader section of society. The old Egyptian Military School was expanded into a Military Academy, fees were reduced or in some cases remitted to enable suitable young men from poorer "lower middle class" backgrounds to enter. These were, however, from the same sort of families from which the *Ikhwan* (*Ikhawan*) Muslim Brethren and Young Egypt Movements recruited most of their followers.

At the same time the standard of education demanded of entrants to this college was raised. All now had to have at least secondary school certificates, which meant a greater level of literacy and general education. However, it also meant that many of the now slightly older entrants arrived with experience of nationalist student politics and, in some cases, involvement in street protests from 1933 to 1936. This seems to have been especially true of those who would graduate from the Egyptian Military Academy (subsequently raised to the status of a College) between 1938 and 1942, who would be a significant group for the political future of their country.

Not that anyone could enter the military. There was still an examination called the *kashf hai'a* which supposedly looked at an applicant's general level of culture and suitability but was actually used to keep out those deemed unworthy of a place in Egypt's armed forces. If it was intended to bar those with suspect political views it failed, although it is also true that many officers who came to see themselves as "revolutionaries" in the late 1940s and 1950s had been simple nationalists and in many cases quite ardent royalists during the first years of their military careers. In the majority of cases, of course, young men volunteered because they were attracted by a military career, especially if they came from prosperous, professional families where there was no particular financial incentive to become an officer.

It is nevertheless wrong to believe that changes in officer recruitment after 1936 immediately and profoundly changed the cultural or ethnic character of the Egyptian officer corps from one that had been Turco-Circassian and almost aristocratic to one that became "native Egyptian" and middle class. The supposed separation between these groups has been exaggerated as almost all sections of Egyptian society, even including parts of the royal family, now identified themselves as Egyptian, whatever their more distant family origins. Meanwhile the pernicious importance of *wasta*, influence and social, family or political contacts, remained.

Hassan Tawfiq and his fellow cadets from the Egyptian Army's Military Academy during a desert training exercise. (Mona Tewfik collection)

A team of cadets from the Egyptian Army's Military Academy, including Hassan Tawfiq [lower right seated] ready for a desert training exercise. They wear the old Egyptian Army's traditional sand-coloured fez covers. The bearded man seated at the front is believed to have been one of the Academy's Muslim chaplains. (Mona Tewfik collection)

What did change was the growing belief that the middle class now had the right to enter the military profession. Of course, young cadets from lower middle-class backgrounds might be teased by more privileged comrades, but almost none could be described as truly poor. Only families with some level of income could afford the secondary education without which entry into the Military College was impossible. The scholar P.J. Vatikiotis summed up these changes very well when he wrote that:

For a long time the officer corps was controlled by the monarch and fairly isolated from the rest of society and its political upheavals … It was the regime's shield against disorder and rebellion … To this extent the officer corps was not, by any standard a vanguard elite in radical Egyptian politics. That is perhaps the reason why when a cadre within the officer corps seized power in July 1952 it did so alone, without the participation or assistance of any civilian organizations that were equally opposed to the ancient regime…

Three of the flying student Hassan Tawfiq's British RAF colleagues "hitching a ride" on a Hawker Audax I Trainer (number K3123) at the RAF's No. 4 FTS at Abu Suwayr around 1936. (Mona Tewfik collection)

The main reasons why most young Egyptians joined the Army in the later 1930s were economic and social. For the great majority a military career offered a steady job and carried increasing prestige, certainly greater than had been the case in earlier decades. Now the appeal of *jundiya* or "soldiering" was spreading beyond the armed forces to embrace the concept of "militancy in the national struggle", an idea seized upon by the quasi-fascist and paramilitary Young Egypt Movement led by Ahmad Husayn.

It is also important to realise that Egypt differed from some other Arab countries in its pattern of military recruitment. In virtually all cases recruitment more or less mirrored the character of local society, and Egyptian society was different from that of its Arab neighbours, not being split by the sorts of religious, tribal and ethnic divisions which bedevilled Syria and Iraq. In the latter cases the officer corps tended to be divided along essentially the same lines as the countries themselves. In the Egyptian armed forces, however, the strongest sense of group loyalty or solidarity was by age, which was itself a reflection of the graduation class to which an individual officer belonged. This continued to be reinforced as promotions generally followed a regular pattern, so that men who graduated together were normally promoted together. Only rarely did this change, as in the case of Hassan Tawfiq whose promotions were delayed by half a year for disciplinary reasons, so that he soon found that his closest colleagues, and thus friends, came from the graduating group which followed his own.

The Sabri brothers were from a more privileged background. The role that the wealthy Sabri family played in Egyptian nationalistic politics during this period highlights the fact that what the British might have called "revolutionary ideas" were not confined to the poor or marginalised. Ali and Husayn Zulfiqar Sabri were part of what became known as "the House of Amin Chamsi Pasha", namely the descendants of a wealthy Egyptian industrialist and politician who had been imprisoned for his support of the Arabi Revolt in 1882. One of his sons, Ali Chamsi (Shamsi), was a co-founder of, and subsequently a dissenter from, the nationalist Wafd Party. A politician like his father, Ali was also Egypt's first representative at the League of Nations.

Two of Amin Chamsi's grandsons by a different son, Abbas Baligh Sabri Bey, then joined the Royal Egyptian Air Force. Of these Husayn Zulfiqar became what his own family later described as 'a disillusioned welterweight prize-fighter and brash World War Two military plane high-jacker' who strove for Sudan's independence prior to briefly being Egypt's de facto foreign policy chief. Somewhat less flamboyantly, Ali Sabri would become a player on the post-1952 Revolution Egyptian political scene, 'poised at the epicentre of power first as prime minister and later as party boss only to be outfoxed by his nemesis Anwar al-Sadat'.

Husayn Zulfiqar Sabri was born in 1915, three years after Amin Chamsi Pasha's death. While he seems to have inherited his grandfather's political passions, he also acquired a passion for fair play and sportsmanship from his British educated father. However, despite his upper class British demeaner and trademark Panama hat, Abbas Baligh Sabri Bey was deeply opposed to British imperialism – especially in Egypt – while also admiring British culture and way of life. Husayn Zulfiqar actually stated in one of his writings that he was greatly influenced by his father and his childhood surroundings, living in an Egyptian home in a very anglicised corner of the leafy Cairo suburb of Ma'adi, almost a "Cairo-on-Thames". When not at school, the Sabri boys spent their time at the Ma'adi Sporting Club where they became known for their sporting prowess.

While his younger brothers earned medals as diving, tennis and water polo champions, Husayn Zulfiqar took up boxing in the welterweight class. Where reading matter was concerned, he had rather unexpected tastes, ranging across French literature, medieval classics and communist tracts. This was the young man whose father wanted him to study engineering in Britain. However, Husayn Zulfiqar dropped out after a year of studies and, rather than return to his disappointed parents in Ma'adi, he tried for a professional boxing career in the United States. There a brief series of victories ended with a serious and sustained hand injury.

Unemployed, perhaps unemployable, and virtually penniless, the proud, hot-headed and relatively dark-skinned young Egyptian became disillusioned with the still racially segregated Land of the Free. Husayn Zulfiqar was particularly bitter because the reality

This remarkable photograph of an Avro Tutor (K3337) in mid-loop should really appear the other way up, as it was photographed by Egyptian flying student Hassan Tawfiq in an accompanying Avro, also in mid-loop. Both machines formed part of the RAF's No. 4 FTS at Abu Suwayr. (Mona Tewfik collection)

of life in the USA proved to have little in common with what he had been told by his family's American neighbours in Ma'adi. Most of them had been missionary-professors working at the new American University in Cairo where they preached tolerance of the "other". In contrast the young Egyptian would-be prize-fighter had found American cities full of discriminatory signs against "blacks" and Jews. With help from one of the Egyptian consulates in America he therefore came home and, at the age of 21 – older than most entrants into the Military College – decided to become an officer, a career choice currently fashionable amongst young Egyptians of his class.

Prince Abbas Halim was president of the Egyptian Gliding Club, the Royal Egyptian Aero Club and the Egyptian Sports Committee. He is seen here at the controls of what appears to be a Slingsby T.3 Primary or Grasshopper, a hair-raising "minimal performance" glider of which the author had experience while at school in north London in the late 1950s. (Nevine Abbas Halim collection)

The overwhelming majority of these young officers volunteered for the Air Force when they graduated, which meant that the Air Force could select the best. Adli al-Shafa'i (el-Shafei) was, for example, the son of a successful doctor and his family had no military tradition. Yet he always wanted to fly and was successful in being accepted by the REAF after graduating from the Military College. Adli al-Shafa'i earned his wings a year and a half later and, like several others who entered the Air Force at the same time, Adli initially served as a fighter pilot before training on multi-engine aeroplanes and eventually serving as a bomber pilot. These young men became a clear and identifiable elite which, in a few unfortunate cases, had a detrimental effect on their behaviour and competence. Nevertheless, in the Egyptian Air Force – as in all others – most focussed primarily on their work and their careers.

Another rather peculiar event during 1936 was when Muhammad Sidqi, who had famously flown his little Klemm L 25 from Germany to Egypt in 1930 (see Volume 4), was briefly unable to renew his pilot's licence (number 347). This was apparently because Sidqi could not prove that he had flown sufficient hours to get his original pilot's licence in Germany. Maybe this event resulted from strains in relations with Germany or perhaps reflected continuing British political interference.

The tense international situation certainly had an impact upon Egyptian politics, especially where the fascistic *Misr al-Fatat* Young Egypt Movement was concerned. Popularly known as the Green Shirts, its executive committee included a number of very conservative and anti-British figures such as General Aziz al-Masri who was the movement's honorary president. General Muhammad Salah al-Harb, the ex-Sanussi military commander

Avro 621 Tutors of RAF No. 4 FTS at Abu Suwayr in Egypt. Because the Tutor was almost identical in many ways to the REAF's Avro 626s, it made the young Egyptian pilots' transition from training to service in an operational squadron very easy. (Mona Tewfik collection)

during the Great War, was also reported to have been an advisor or supporter.

Even Prince Abbas Halim briefly flirted with *Misr al-Fatat* in the mid-1930s. In 1936, however, the flying prince tried to organise a strike at the Misr Spinning and Weaving Company in al-Mahalla al-Kubra. When this failed, Abbas Halim dropped out of politics for a while, reappearing the following year when he attempted to take over the leadership of the Egyptian "Committee to Organize the Workers' Movement". A short time later he tried to lead the Cairo Tramway Workers' Union and helped organise a Joint Transport Federation, none of which proved very effective. Prince Halim was from a generation which saw the Egyptian trades union movement as part of a broader nationalist struggle against British imperialism. Meanwhile European concepts of "class struggle" were taking root amongst factory workers along with the Egyptian Communist Party, but not amongst the rural poor, and Prince Abbas Halim's influence declined along with that of the old nationalist trade unions.

Muhammad Sidqi Mahmud al-Milaighi (middle row far left) and another Egyptian student pilot (middle row far right) in E Flight, No. 32 Course of the RAF's No. 4 FTS Abu Suwayr from July 1935 to April 1936. This was the first time that Egyptian military pilots would be trained in Egypt. Sidqi went on to have a distinguished career in the REAF, EAF and UARAF, until he was unfairly blamed for the latter being almost wiped out on the ground by Israeli air strikes in June 1967. (Squadron Leader C. Wright-photo)

Negotiations to purchase new and more powerful aeroplanes had started before the new Anglo-Egyptian Treaty was signed. The machines of interest included an Avro 652 Mk. II which initially carried the Egyptian civil registration SU-AAO (c/n 891) and was flown to Cairo to be tested for a proposed new bomber squadron. It was then given the military serial number W204. This civilian machine looked very similar to the Anson and is often called the Anson prototype. In fact, it was a small transport aeroplane capable of carrying four passengers plus a crew of two. It also had a more sloping windscreen, a rounded nose, a door on the port rather than the starboard side of the fuselage and, as a civilian aeroplane, lacked a gun turret. It was powered by two Armstrong Siddeley Cheetah IX rather than Cheetah V engines as on the Anson Mk. I. The Egyptian tests were a success and in October and November three Anson Mk. Is were delivered, becoming W205 to W207. Egypt subsequently received three more ex-RAF machines.

On 22 December 1936 the new Anglo-Egyptian Treaty was ratified by both parties and came into full effect. That month also saw the purchase of a squadron of Audaxes, the first machines actually reaching Egypt in 1937. This "Egyptian Audax", as the type was popularly known, was a radial engine version of the basic Audax, built under licence by Avro at Newton Heath near Manchester and designated as the Avro 674. For Egyptian aircrew who by now had considerable experience on the Avro 626, the transition to "Egyptian Audaxes" was relatively straightforward. Nevertheless, the latter were considerably more powerful and, in preparation for their arrival, the pilots practised aerial combat manoeuvres that they had never been taught before, along with aerobatics and bombing.

The plan was for the Egyptian Air Force to have two general-purpose squadrons by the end of 1937, plus communications and training squadrons, all under the direction of an Air Force

Muhammad Sidqi Mahmud al-Milaighi second from the left in the rear file of E Flight, during the passing out parade of No. 32 Course, at the RAF's No. 4 FTS Abu Suwayr, April 1936. He is distinguished by his red fez, slightly darker uniform and lack of puttees. (Squadron Leader C. Wright photograph)

Avro 626s of the REAF arrayed on the apron at Almaza aerodrome around 1936. The hangar displaying the green and white roundel of the Egyptian Air Force would appear in several photographs during this period. (Author's collection)

of 'almost proto-dynastic antiquity'. Furthermore, the Military Staff and High Command and Staff were 'perhaps the weakest component in the whole of their military fabric'. This was, of course, before the arrival of a substantial number of new junior officers who graduated in and after 1937.

Despite widespread patriotic enthusiasm for stronger armed forces, strains were already being felt, not least because of a theoretically voluntary levy called the National Defence Fund. Organised in February 1937, this was resented by the poor, many of whom felt browbeaten or bullied into handing over money. Other problems resulted from the fact that there were simply not enough Egyptian officers of sufficient rank and experience to take on some of the roles demanded by a sudden military expansion. This was particularly acute in the EAAF which was at the time still part of the Egyptian Army. Here it had originally been assumed that senior officers like Tait and others seconded from the RAF would lose their jobs and be replaced by Egyptians. Nevertheless, it rapidly became apparent that such an ambition was impractical as there were no Egyptian Air Force officers of higher rank than *Yuzbashi* (Captain). Thus, the first Half Yearly Report by the British Advisory Mission attached to the Egyptian Force dated 26 April 1937, recorded that all three squadrons were still commanded by British officers, though Nos. 1 and 2 Squadron did now have Egyptian seconds in command.

Headquarters. But before this could be achieved, British personnel had to leave and, on 6 January 1937, a ceremonial dinner marked the departure of all British officers from the Egyptian Army. Their role had been to command and to advise, and for the latter task they would be replaced by a permanent British Advisory Mission. This, as its name indicated, could only advise, not command.

As British officers attached to the Egyptian Army left during January, the new British Military Mission wrote a candid report on the force which the British had supervised and dominated for half a century. It was far from flattering. There were no machine guns, no support, no anti-tank weapons, inadequate or downright inferior artillery, no tanks, no wireless and no anti-aircraft guns. Military training was, in the words of this report, based on tactics

Also in April, the Egyptian Army Air Force (EAAF) became a separate service called the Egyptian *Silah al-Tayaran al-Maliki*. This is generally translated as the Royal [Egyptian] Air Force or REAF, though Royal Flying Forces might be a more accurate translation. It was not until May 1947 that the Egyptian Air Force was again renamed as *Al-Silah al-Jawi al-Maliki* which does translate as "The Royal Air Force". In neither case was it thought necessary to insert the word *Misri* or "Egyptian", though foreigners invariably referred to the Egyptian Air Force as the REAF.

Early in 1937 the new Egyptian Commander of the REAF was ordered by his government to establish a High Institute for Aviation at Almaza aerodrome, to emphasise the REAF's somewhat nominal autonomy from the RAF. Its students would also use a small landing ground near Khanka. In December the REAF's new Flying Training School (FTS) started teaching its first "cohort" of five students. This Egyptian FTS would offer elementary and advanced flying training, firmly based upon a model offered by the RAF's No. 4 FTS at Abu Suwayr and, more distantly, by the RAF College at Cranwell in England. Some of the REAF's older DH 60 Gipsy Moths were transferred to this new FTS as primary trainers.

The observer-gunner climbing into the rear cockpit of an REAF Avro 626 at Almaza around 1937. (Author's collection)

The school itself was staffed by the Air Force's most experienced pilots after they received further specialised training in Britain. Ibrahim Hakki, for example, found himself attached to the FTS as navigational instructor. Some Avro 626s were also now transferred to the school as intermediate trainers. However, as the venture proved successful and further flying cadets were enrolled, it became clear that new training aeroplanes would be needed. A civilian flying school had been operated by Misr Airwork at Almaza since 1936 and there is evidence that the two establishments collaborated, while it is also possible that some of the Air Force's Gipsy Moths may eventually have been handed over to the Misr Airwork Flying School, though this is unconfirmed.

RAF and Egyptian REAF flying students fooling around an Avro 504K that had tipped onto its nose at No. 4 FTS Abu Suwayr in the mid-1930s. (Mona Tewfik collection)

As in almost all air forces, the REAF selected some of its most skilled pilots for further training as instructors. One such was Muhammad Hassan al-Maghribi who had entered the Egyptian Military College in 1932. After graduating in the Class of 1935 and being selected for the REAF, he was sent to the British No. 4 FTS at Abu Suwayr where he qualified as a pilot in 1936. Two years later al-Maghribi was sent to England, to train as a flying instructor in the RAF's Central Flying School at Upavon. There he was also awarded a Flight Engineer's Certificate on single and twin-engine aeroplanes. Muhammad Hassan al-Maghribi's appointment as a flying instructor at Almaza in 1939 was followed in 1940 by the British Ministry of Aviation awarding him a Grade One "flight characteristics teacher" and a qualification as a "teacher of aviation instructors". Muhammad Hassan al-Maghribi would, in fact, eventually become the best-known instructor in the history of the REAF.

Meanwhile a temporary relaxation in international tensions enabled the Royal Egyptian Aero Club to hold the delayed Tour of the Oases Air Rally in February 1937. It was supposedly organised by the Aero Club, which thereafter received credit for its success, but in practice most organisation and planning had been done by what was currently the Egyptian Army Air Force. The EAAF's hangars were almost emptied to accommodate the Rally's numerous competitors and the first take-offs were to start in the morning of Tuesday 23 February, the machines leaving Almaza aerodrome at specified intervals.

Their route was ambitious, initially taking them eastward to the Gulf of Suez, then down the western shore of the Red Sea, inland at Hurghadah, then south along the River Nile to Aswan – a total of about 800km. The following morning would be one of relaxation, visiting archaeological sites, before the aeroplanes started their scheduled take-offs at 2:00 p.m., heading northwards to Luxor. There would be no flying the next day, Thursday, which would

The British Aircraft Swallow was a licence-built version of the German Klemm L.25, this example having been flown by Miss Lillie Dillon during the 1937 Circuit of the Oases Air Rally. The tail skid broke when Miss Dillon landed in a ploughed field at Aswan, but was repaired with assistance from *Bimbashi* Stocks of what was still the EAAF. (Author's collection)

again be spent sightseeing. The most strenuous and potentially most dangerous day would be Friday 26th, with the first early morning take-offs scheduled for 5:45 a.m. The aeroplanes would then fly around the Western Desert oases of Kharga, Dakhla, Farafra and Bahariya, before heading north-east to Almaza. The total distance flying was about 1,100km.

The competitors came from many countries and the Tour of the Oases Air Rally was intended to test the endurance and reliability of aeroplanes rather than the skills of pilots, so the rules stated that spare parts were not allowed. Damaged elements could, as far as it was possible, be repaired with the help of the EAAF which stationed men and equipment at all landing points. For example, when the tail skid of Miss Lillie Dillon's British Aircraft Swallow (licence-built Klemm L 25) broke at Aswan, *Bimbashi* Stocks helped repair it. *Qa'im Maqam* Webster followed behind the rally in the EAAF's new Avro 652, ready to find any lost aeroplanes. None were and no participant was missing for more than an hour or so. Meanwhile the EAAF's Westland Wessex was based at Dakhla Oasis in case of need. When the Tour of the Oases Rally was over, Dr Abu Zahra of the Royal Egyptian Aero Club gave fulsome thanks to the Egyptian Air Force for its help during this stressful exercise.

On 4 March 1937 a report in the British magazine *Flight* demonstrated the sunny outlook of aviators at that time, before the dark clouds of a looming war:

Round the Oases. Opening Stages of the Egyptian Royal Aero Club's Meeting Described by One on the Spot: Nine British Participants (Cairo, Tuesday, February 23).

Forty-one competitors started this morning on the Circuit of the Oases, and that in itself is an eloquent testimony to the high standard of organisation of the Royal Aero Club of Egypt. Only two are left behind, one on account of some hitch in insurance and the other with mechanical trouble.

The weather is now perfect and there is every promise that the next four days will pass off as happily for everyone concerned as did the start this morning.

First Arrivals

The first arrival reached Almaza Aerodrome on Friday of last week and since then the aerodrome has worn a very gay appearance. A large aircraft park is laid out on the north side of the big hanger and between the hanger and Misr Airwork clubhouse a reception tent has been put up.

Flight magazine then gave a list of competitors, some of whom were flying what are now considered "rare birds":

Austria (1): Otto Mendl (Messerschmidt 108).
Belgium (1): J.M. Provost (Leopard Moth).
Czechoslovakia (4): Kutloch (Baby Okypn; Zlinska Letecka A.S. (Ambrus) (Zlin XII); Zlinska Letecka A.S. (Fukfu) (Zlin XII), Aeroklub de Tchecoslovique (Benes Barz) (BF 51).

Egypt (5): Ahmed Salem (Leopard Moth); Mlle Lotfia el Nadi (Hornet Moth); Prince Omar Halim (Miles Hawk Major); Prince Abbas Halim (Waco YKS); A. Nagi (Avro 641 [an EAAF pilot in Tait's Commodore]).

England (9): F.C.J. Bulter (Hornet Moth); E.D. Spratt (Miles Falcon); Major J.C. Hargreaves (B.A. Eagle); V.A.P. Budge (Miles Hawk Major); Capt. J.K. Mathew (Miles Hawk Major); Fl. Lt. J. Heber Perry (Puss Moth); P.G. Aldrich-Blake (Percival Vega Gull); Misses M. and S. Glass (D.H. Moth); Miss Lillie Dillon (B.A. Swallow).

France (11): H. Lumière (Caudron 635); G. Morizot (Caudron Aiglon), H. Engernich (Leopard Moth); E. Wattine (Caudron 635); A. Masset (Caudron Aiglon); Mme G. le Pelly du Manon (Caudron Aiglon); De Chateaubrun (Percival Vega Gull); Marquis de Saurez D'Aulun (Caudron Aiglon); Hirsch Ollendorf (Vega Gull); Guy Hansez (Caudron Simoun); A. Boulanger (Farman 403).

Germany (4): Otto R. Thomsen (B.F.W Me 108b); President von Gronon [Gronow?] (B.F.W. Me 108b); Freih. Speck von Sternburg (Junkers Ju.86); Karl Schwabe (Klemm Kl 32a).

Poland (1): J. Drzewiecki (R.W.D. 13).
Rumania (1): Prince Bibesco (Potez).
Syria (1): H. Gagey (Farman 402).
Italy (3): G. Zapetta (F.N. 305); V. Biani (Ghibi [probably Caproni Ca.309 Ghibli]); S. Santini (Ghibi [probably Caproni Ca.309 Ghibli]).

The De Havilland DH 87A Hornet Moth (registration SU-ABT) flown by Lutfia al-Nadi during the Circuit of the Oases Air Rally 1937, with its large rally number 49. Lutfia al-Nadi was a pioneer aviatrix in Egypt and would play a significant role in making her fellow Egyptians air-minded. (Author's collection)

A Praga E.114M Air Baby (registration OK-IPN, rally number 59) photographed on 1 March 1937 during the Tour of the Oases Rally. It was flown by Mr Kutloch. The aeroplanes were guarded by men of the REAF. (Author's collection)

Another aeroplane which took part in the Circuit of the Oases Air Rally early in 1937 was this Avro 641 Commodore (registration SU-AAS, rally number 41) which belonged to Victor Hubert Tait, commander of the EAAF. It was later transferred to what had become the REAF and was given the military serial number W203. (V.H. Tait collection)

During the Second World War it was rumoured that the German participants included senior Luftwaffe pilots who used the rally to look for possible military landing grounds. While Otto R. Thomsen and Karl Schwabe were leading German "sports pilots", Freiherr Speck Von Sternburg appears to have been the same Major Manfred Freiherr Speck Von Sternburg who, born in 1904, was killed on 26 October 1940 while commanding the Heinkel 111 bombers of the Kampfgeschwader 27 III. Gruppe. For the Tour of the Oases Rally his Junkers Ju-86 (German civilian registration D-AKOP) crew included Hauptmann (Captain) Von Bloomberg, Hauptmann (Captain) Von Salomon and Funker (Radio Operator) Posner. Their aeroplane was a Ju-86 C-1 (c/n

The biggest machine to take part in the 1937 Circuit of the Oases Air Rally was a diesel-powered Junkers Ju-86 (registration D-AKOP, race number 3) named *Kismet*. Flown by Freiherr Speck von Sternburg and his crew, *Kismet* was declared the Rally's overall winner, followed by another German pilot, Otto Thomsen in a BFW Me-108b, while Guy Hansez of France came third in his Caudron Simoun. (V.H. Tait collection)

Prince Abbas Halim's own Waco UIC (registration SU-AAN) during the Circuit of the Oases Air Rally in 1937, when it had rally number 22. (V.H. Tait collection)

860218) powered by two Junkers Jumo 205C diesel engines and owned by the Junkers Company. Six of this transport version of the better-known bomber would be constructed for Deutsche Luft Hansa.

Flight continued its detailed account:

There are, of course, two events: First, the Circuit of the Oases – in no way a speed contest – and; finally, the Speed Race.

There followed a detailed explanation of the complex formulae whereby marks were awarded. The speed competition carried no handicap and was open to all aircraft which completed the Oases Circuit, focussing on a 104-mile (167.4km) triangular part of the course:

During the six days the competitors are in Egypt, they are the guests of the Royal Aero Club. Yesterday evening February 22, H.E. Mohamed Taher Pasha [Muhammad Tahir Pasha, president of the Royal Aero Club of Egypt and later Chairman of the Egyptian Olympics Committee from 1946 to 1954] gave a cocktail party at Heliopolis Palace Hotel to the competitors in the Rally and to people connected with aviation in Egypt.

The aircraft park during the last two days has excited a lot of interest. From the point of view of private owners, the Messerschmitt and the Vega Gull are perhaps outstanding. In regards the larger machines, the Junkers 86, with two Jumo motors is the centre of attention. Oddly enough, a second Junkers 86 arrived yesterday and took off this morning for Tehran and Kabul [this flight proved to be of considerable interest to the British diplomatic corps in Cairo].

One of the Italian entries, the F.N. 305, is a very neat affair and is expected to prove one of the fastest machines in the Rally if she shows her paces in the speed race.

Flight's report concluded with an expression of thanks and congratulations to the organisers of the Air Rally:

Before ending this preliminary note, mention must be made of the excellence of the organisation. The Royal Aero Club of Egypt, and in particular of the President, H.E. Mohamed Taher Pasha and the secretary, Dr M.A. Zahra, both of whom have been indefatigable in their efforts to make the meeting a success, are to be warmly congratulated. To get almost forty aircraft of eleven different nationalities serviced and started without a hitch is no mean undertaking. The notes with which all competitors have been supplied is probably one of the best of its kind produced for such a race. The route instructions and information given are clear and complete to the smallest detail.

The overall winner of the Tour of the Oases Rally was Freiherr Speck Von Sternburg in the Junkers Ju-86, Otto Thomsen being second in a BFW Me 108b while third was Guy Hansez of France in a Caudron Simoun. *Flight* ended its report with a brief update:

Since the above was received, the result of the speed race has become known. Our correspondent's anticipation was well founded, for the Italian F.N. 305 proved the winner, M. Guy Hansez was second and the only British competitor (in the speed race), Mr. S.B. Cliff, retired. – [Ed.] Guy Hansez was flying one of the Caudron Simouns. S.B. Cliff was not in the list of competitors given above.

There had, in fact, been seven participants in the speed race, but only four finished. Giovanni Zapetta came first, Guy Hansez second, Henri Lumière third and Otto Thomsen fourth. The poor performance by Thomsen's Messerschmitt 108 in the speed race came as a surprise to many and is said to have deeply annoyed both Hitler and Hermann Goering.

A group of interested EAAF officers was photographed with one of the Caudron Simouns, probably that registered F-ANXB

REAF officers standing next to the Caudron Simoun flown by Guy Hansez during Egypt's Circuit of the Oases Rally in 1937. (V.H. Tait archive)

Amir Alay (Col.) Ali Islam Pasha, the commanding officer of the REAF from 1936 to 1939. As a senior Egyptian Army engineer officer, he had previously headed the Egyptian Military School. (Author's collection)

Flight Lieutenant G.B. Keily, an instructor at the RAF's No. 4 FTS Abu Suwayr in the 1930s, taught a number of EAAF/REAF pilots including Hassan Tawfiq. Shot down by Italian fighters over North Africa on 18 September 1940, Squadron Leader Keily, as he was by then, became a prisoner of war. He retired as an Air Commodore in 1955. (Mona Tewfik collection)

Wing Commander S.N. Webster, Squadron Leader D.W. Reid, Squadron Leader P.B. Coote and Squadron Leader C.E.N. Guest either joined or remained with the RAF Advisory Mission. Four were attached to the REAF's squadrons where they were primarily concerned with training. There were also 11 RAF Warrant Officers and NCOs attached to the REAF as advisors involved in training and technical support.

These men were and owned by the manufacturers, SA des Avions Caudron/Issy. This was the machine flown by Guy Hansez and had the race number 4. *Amir Alay* Tait's Avro 641 still had the civil registration SU-AAS (ex- G-ACRX) when it took part flown by Ahmad Nagi, and had not yet been transferred to the Egyptian Force. In fact, no military aircraft were allowed to take part in the Rally.

Only a few weeks after the success of the Tour of the Oases Air Rally, the EAAF became the REAF and got a new Commanding Officer, *Amir Alay* (Colonel) Ali Islam Bey, a senior Egyptian Army engineer officer who had previously been at the head of the Egyptian Military School. He remained at the head of the newly formed REAF until May or August 1939. The contracts of the British and British Imperial officers who had been seconded to the EAAF came to an end on 7 January, but they remained in position pending new contracts as part of the British Military Mission under Major General James Handyside Marshall-Cornwall.

The EAAF/REAF had made clear that they were not happy about the prospect of losing their RAF colleagues, whatever Egyptian politicians might say, and they practically insisted that Tait remain in some capacity. As a result, not only did Group Captain V.H. Tait become Air Advisor to the Ministry of War in Cairo (he would be awarded the OBE for his services in 1938), but Wing Commander N.P. Dixon became Staff Officer to the REAF's new Egyptian CO. Meanwhile Wing Commander E.A.C. Britton,

undoubtedly welcome as, according to a report to the British Foreign Office in London from the British Ambassador Miles Lampson, Egyptian pilots threatened to refuse to leave the ground if British NCOs were removed. This was reportedly because there were not yet sufficient Egyptian riggers and fitters capable of functioning without direct supervision. Lampson was probably exaggerating, but there is no doubt that the REAF still needed maintenance and technical support from the RAF. In fact, within four weeks of the withdrawal of the original RAF training team, there had been six accidents. Most were minor but one pilot was killed when his Gipsy Moth crashed.

Officers of the REAF were selected from the elite of the Egyptian Army's Military College but went through the same harsh desert warfare courses as those who would become Army officers. The cadets seen here include at least two who would enter the Air Force: Hassan Mahmud seated in front with bushy eyebrows, and Hassan Tawfiq kneeling behind his right shoulder wearing a fez with the Egyptian Army's distinctive khaki cover. (Mona Tewfik collection)

Headquarters, and there is evidence that the REAF almost immediately began to suffer from the political interference which would hamper its operations for decades to come. Meanwhile some officers clearly resented being under the command of an Army officer, despite the fact that everyone, Egyptian and British alike, agreed that *Amir Alay* Ali Islam Bey was a thoroughly likeable if rather easy-going man.

The career of Hassan Tawfiq is one of the few that is known in detail during this period. Following his foolish crash at Sinbillawain on 13 June 1936 (see Volume 4) and the consequent six months delay in his promotion, the young Hassan Tawfiq did not officially become a *Tayar Thani* ("Second Pilot", or Pilot Officer in RAF terms) until early 1937. He was then sent from the REAF's new FTS to his first squadron. There Tawfiq served for only a short period as an operational pilot, partly because of his injury and partly because his evident engineering skills made him more useful in the mechanical, maintenance, wireless and eventually radar aspects of operations. For this reason, Hassan Tawfiq would be sent for further advanced training in Britain in 1938.

The year 1937 saw a number of younger men enter the Military School (soon to be raised to the status of an Academy) who later became members of the Free Officers Movement. They included Gamal (Jamal) Abd al Nasser who enrolled on 17 March 1937, Anwar Sadat, Hassan Izzat and Wagih Abaza. While the former two remained in the Army and eventually became Presidents of Egypt, the latter two joined the Air Force. All four were, according to Egyptian Police files, believed to be members of either the *Misr al-Fatat* Young Egypt Movement or the *Ikhawan* Muslim Brotherhood. These organisations may, in fact, have used their influence to help young men get into the Military School. Some would nevertheless graduate with the best grades, Nasser himself having been selected as one of the few entrants who took the intensive one-year course, instead of the Military School's usual three-year course. Others included Abd al-Latif Baghdadi and Jamal Mustafa Salim, who were part of the cream of the crop selected for the REAF. According to some sources, both were at one time either members of, or had close associations with, the Muslim Brotherhood.

A letter written by Abd al-Latif Baghdadi to his brother Sa'ad, dated 1 November 1937, was published many years later. In it he described his feelings at being accepted by the Military School: 'I begin my letter in the name of God. God made me proud of myself and proud of my conviction that he helped me get accepted by the Military College. My joy was beyond price because I had longed to [attend] that school but I did not know the reason why, maybe because I loved the military life and because I wanted to serve my dear homeland and give my life for it, along with everything that was in my hands'.

Hassan Tawfiq at the RAF's No. 4 FTS in Egypt, perhaps wondering whether he was going to need one of those parachutes. (Mona Tewfik collection)

Another report noted that, by 26 April 1937, the Egyptian Air Force still only consisted of 27 officers, plus the three British officers directly attached to the squadrons, while its total personnel including technicians was only 415 men. The great majority worked at Almaza, which was the REAF's Operational

The formal graduation photograph of a cohort of the Egyptian Army Military Academy cadets on parade in Cairo, 1938. Amongst them was newly commissioned Muhammad Abd al-Hamid Abu Zaid who then went straight into the REAF, along with a select group of the Academy's best students. (Abu Zaid family collection)

Another youngster from this generation of cadets and future revolutionaries was Hassan Ibrahim who was certainly a member of the Young Egypt Movement in Alexandria. Meanwhile *Amir Alay* Ali Islam Bey decided that the REAF needed every qualified man available, and so recalled Abd al-Hamid al-Dighaydi from his punishment posting in the Army. Tait thought it a bad idea and said so, telling Ali Islam face to face, 'I warn you about this'. Thereafter al-Dighaydi appears to have flown anti-drug smuggling patrols over Sinai before he was appointed Officer Commanding the REAF's Flying Training School at Almaza. Al-Dighaydi was, however, replaced early in June 1939 by Nu'man Nada. By then Ali Islam Bey had retired in May 1939, so it is possible that Abd al-Hamid al-Dighaydi lost a patron and protector. He would find himself in trouble again during the Second World War, being transferred back to the infantry – this time permanently. Nevertheless al-Dighaydi would win renown as a hero of the Faluja Pocket during the Palestine War 1948–1949.

Back in 1937, the Egyptian Army not only lacked proper weaponry but also support and training facilities. Furthermore, it was accepted that the Air Force would remain in practice, if not in theory, under the operational control of British officers who were themselves under senior Egyptian officers. Thus, at the start of the year, the EAAF's 38 machines were formed into Flights rather than full squadrons, these being two general service Flights, one training Flight and one communications Flight. The machines were normally unarmed and no real bombing practice had yet been undertaken. The expansion of the Egyptian Air Force also put significant strain on the country's finances. According to *The Aeroplane* magazine, by 6 January 1937 Egypt had spent £300,000 on British aeroplanes – a considerable sum in those days – which would bring the Egyptian Air Force's strength up to 49 machines from its current 38. Even as they stood, such plans were expected to cost around half a million pounds per annum for the next three years to train personnel and expand the force to 78 aeroplanes in four squadrons, rising to 100 by 1941.

The coming of the Second World War would of course upset these plans, but what the Egyptian authorities had in mind remained a small but modern mechanised Army, supported by an effective Air Force. The latter should consist of a training squadron, an army cooperation squadron and a bomber-transport squadron based at Almaza, plus a bomber squadron and perhaps a fighter unit in Alexandria. Expansion seemed to proceed apace so that by the end of April, when the EAAF officially became the REAF, its strength stood at six newly delivered Avro 674 "Egyptian Audaxes", 22 Avro 626s (including four more replacement machines, serial numbers J330, J330, J331 and J332), six elderly De Havilland DH 60 Gipsy Moths IIs, one Avro 652, one Westland Wessex and an Avro 641 Commodore. One of the replacement Avro 626s, number J330, appears to have duplicated the serial number of an earlier aeroplane, unless the latter had been completely rebuilt.

The three squadrons were No. 1 (Army Cooperation) with nine Avro 626s and three Audaxes whose role was to support army units stationed on the frontiers, No. 2 Squadron which had the same complement of aeroplanes as No. 1, and No. 3 at Almaza which currently had six Moths and the four Avro 626s used by the FTS, plus the Avro 652, the Westland Wessex and the Avro Commodore for communications duties. All three squadrons had British officers in command, with an Egyptian officer as second-in-command though their positions would eventually be reversed.

Of the new aeroplanes, the Westland Wessex had been flown out of Heston in west London on 15 March, piloted by Mr D.P. Cameron. It reached Egypt by the now usual route via Italian Libya, on 21 March. The Panther VIA engine Avro 674 "Egyptian Audaxes" had broad, low pressure "balloon" tyres which should perhaps better be called sand tyres. The arrival of the first batch of these more powerful machines (serial numbers K400–K405, c/n 943 to 948) late in 1936 had a very positive impact upon REAF morale, and was followed a few months later by six more. Sometime in April 1937, *Amir Alay* Tait went to the United Kingdom to purchase a reported "two squadrons' worth" of the Panther X powered version of the "Egyptian Audax", to be delivered in 1938 (serial numbers K501–K518, c/n 1035 to 1052).

According to some sources, the REAF – or perhaps King Faruq himself – purchased a Fleet Model 7 Fawn in or around 1937, which was supposedly given the serial number F1 as the first machine of a proposed Royal Flight. The Fleet Fawn was a two-seater primary trainer, designed by Reuben Fleet and built in Canada. It entered service with the Royal Canadian Air Force in 1931, but there is no evidence that such a machine ever reached Egypt. Certainly, no more were ordered and the entire story is probably a myth or misunderstanding.

In 1937 Armstrong Siddeley, the British manufacturer of motor cars, aero-engines and more recently aircraft, reportedly offered to establish a factory to assemble aeroplanes in Egypt, but the

Hawker Audax I number K3113 of the RAF No. 208 Squadron crashed on 10 May 1937 in Egypt with the loss of its occupants. The wreck is guarded by local policeman. (Albert Grandolini collection)

offer was declined. In fact, John Davenport Siddeley's interests in the aviation side of this business had already been purchased by Tommy Sopwith, the famous aeroplane designer and current owner of the Hawker Aircraft Company. This therefore became Hawker Siddeley, a company perhaps best known for the Hawker Hurricane fighter of the Second World War. In fact, the reality behind Armstrong Siddeley's initial offer to Egypt remains a bit of a mystery, though it would be repeated two years later.

On 18 April 1937, while on a visit to France, King (though as yet uncrowned) Faruq visited the village of Monchaux-Soreng near Blangy-sur-Bresle in Normandy, to see where his cousin Fu'ad Abd al-Hamid Haggag had been killed during the first Avro 626 delivery flight. While the King was there, the Egyptian ambassador in Paris and the Conseiller General of the canton of Blangy agreed that a monument would be built, but because of the Second World War this never happened. On a more positive note, Britain agreed to sponsor Egypt's membership in the League of Nations, the country being accepted on 26 May 1937.

On 29 July 1937 the coronation of King Faruq resulted in a number of seemingly symbolic changes which proved to have greater significance than might have been expected. For example, from August onwards Egyptian military officers took their oaths of loyalty directly to the King rather than to the country or its constitution. This was, of course, also the case in Britain and its empire so the British were hardly in a position to complain. Thereafter King Faruq not only used every opportunity to stress that the Army and REAF were separate from and thus "above" politics, but also to focus their loyalty upon himself – to make them "his" armed forces. Hassan Sabri Pasha, the Egyptian Minister of Defence, then resigned when officers were not given greater privileges than civil servants, maintaining that the traditional status of the Army or *taqlid al-jaysh* had not been respected. In the backs of the minds of many people was, of course, the military coup in Iraq of 29 October 1936.

On 3 May 1937 No. 4 Squadron of what was now the REAF moved from Almaza to the civilian airfield at Dakhailah outside Alexandria. There its personnel had to live and work in tents, using part of the nearby civilian hangar for aircraft maintenance. Dakhailah would remain No. 4 Squadron's base for reconnaissance in the Western Desert and along the Libyan frontier until a proper airfield was built at Marsa Matruh. Here work was currently delayed because of a lack of funds, hence aeroplanes operating from Marsa Matruh had to do so from the beach. Existing airfields were also expanded and work started on some new ones in the Suez Canal area. Work clearing and constructing airfields would continue with greater urgency in 1939, several of those in the Western Desert becoming well known as bases of the RAF's Desert Air Force during the Second World War.

In December, No. 1 Squadron REAF's colleagues, No. 208 Squadron RAF, returned from Palestine to Heliopolis where they remained for some time as the main British Army Cooperation Squadron in Egypt. Meanwhile REAF training continued with increasing intensity. Amongst those who were chosen for additional instruction in Britain was Salih Mahmud Salih. Not yet having qualified as a pilot, he apparently still held the rank of *Mulazim*, Lieutenant, though with a total of 52 flying hours, when he was attached to No. 4 (Army Cooperation) Squadron RAF at Odiham in Hampshire in 1937. While there Salih Mahmud Salih was lucky to escape with his life from a serious crash on 6 July (in RAF Hawker Audax, serial number K2026). Having recovered from his injuries, he received his wings in December 1937 with a total of 830 hours.

Elementary flying training continued at Almaza where the FTS staff included four RAF officers on secondment to the Egyptians as instructors. They were one Squadron Leader, two Flight Lieutenants and one Sergeant Pilot. According to a report dated 20 December 1937, the first batch of ten Miles M.14 Magisters had arrived, with others due to be delivered soon. Some sources state that the first order was for 23 Magisters though others suggest it was 36. The REAF's first Magisters were given serial numbers L201 to L210, eventually running to L243. The Miles M.14 Magister, also known as the Hawk Trainer Mk. III, was developed from the highly successful civilian Hawk Major or Hawk Trainer. The only significant difference was that the military versions had enlarged cockpits to accommodate parachutes, plus full blind-flying equipment. Racks for eight small bombs could be added beneath the centre-section. The M.14 was accepted by the RAF despite the British Ministry of Defence policy of only wanting metal aircraft. It was also the RAF's first low-wing monoplane primary trainer.

A school to train REAF mechanics was formed at Almaza, with an RAF Sergeant Fitter as instructor. Its first 90 recruits came from various trade and technical schools in Egypt, and thus had some grounding in the required skills. A few months later authority was given for the establishment of an Armament and Wireless School which was expected to accept 48 pupils at a time and had an RAF Sergeant Fitter-Armourer attached as chief instructor.

The last full year of peace before the outbreak of the Second World War was not without controversy in Egypt. The amiable *Amir Alay* Ali Islam Bey continued as Director or senior officer of the REAF under the Chief of Staff of the Egyptian Army, though

Shiny and new Miles M.14 Magister primary trainers lined up at Almaza shortly after their delivery to the REAF in 1937. They were about to be inspected by Egyptian Minister of Defence, Salih Harb Pasha. (Author's collection)

This Miles M.14 has what appears to be the Circuit of the Oases number 5 pasted onto its rear fuselage. However there also appears to be an REAF roundel behind the unidentified pilots, and no military aircraft took part in the 1937 Air Rally. In fact the 5 might be an early example of the application of the final digit of REAF primary training aircraft, in this case L205. (V.H. Tait collection)

the Egyptian Prime Minister, Muhammad Mahmud Pasha (in office 29 December 1937 to 18 August 1939) returned from a visit to London and promptly appointed General Aziz al-Masri as the Army's new Inspector General. Apparently, General al-Masri travelled to Germany in 1938 and, according to some hostile British accounts, did so as an "emissary of anti-British circles, hoping for supplies of arms". It seems highly unlikely that this was the real – and certainly not the official – reason for the visit. In fact, Aziz al-Masri did not remain in this post for long, being moved to the position of Chief of Staff in 1939. Muhammad Sidqi's friend from when he learned to fly in Germany, the First World War "ace" and currently the Luftwaffe's Director General of Equipment, Ernst Udet also visited Libya and Italian East Africa in January 1939. He made several flights around these areas, including one along the Kenyan frontier, much to the concern of the British authorities.

King Faruq's marriage to a young Egyptian woman on 20 January 1938 proved very popular, his bride being Safinaz Zulficar, a member of the Egyptian aristocracy of Turkish origin whose father was a highly respected judge. On becoming Queen, Safinaz Zulficar took the name Farida. A few months later the King was praised for a speech from the

in practice Muhammad Abd al-Muna'im al-Miqaati ran the REAF under Ali Islam's somewhat nominal command. In July 1938

The REAF's second delivery of Hawker "Egyptian" Audaxes (including number K510) drawn up on the apron at Almaza aerodrome outside Cairo, facing a line-up of earlier Avro 626s. Delivered in 1938, these Audaxes were powered by more powerful Panther X engines. (EAF Museum collection)

A classic photograph of an REAF aeroplane in flight over Gezira Island in Cairo. Miles Magister serial number L204 has the final digit 4 repeated very large on the rear fuselage. (EAF Museum collection)

throne on 13 April, in which he stressed the need to strengthen the Egyptian Army and improve its weaponry – a popular theme in a country which had recently regained its independence from the British. In fact, the money allocated by the Egyptian government for defence was roughly doubled.

Amongst the professionals of both British and Egyptian armed services, relations were good and, according to what would become regular reports by the British Advisory Mission, Major General Marshall-Cornwall wrote on 24 May 1938 that the REAF was making 'remarkably good progress and must now be regarded as a definitive factor in the forces available for the defence of Egypt'. The structure of the REAF currently consisted of the HQ Air Section of the Ministry of War and Marine, the Station HQ at Almaza, No. 1 (General Purpose) Squadron with Avro 626s and Audaxes, No. 2 (General Purpose) similarly with Avro 626s and Audaxes, No. 3 (Communication/Bomber) with the Avro 652 and Westland Wessex, the Training Squadron with Gipsy Moths and Magisters, and a small RAF Mission to assist in training. The British Advisory Mission report of 24 May also stated that on 30 April 1938 the REAF's strength numbered 61 aeroplanes, 853 Egyptian personnel, and 20 British advisors. During May 1938 an REAF football (soccer) team took on a team from RAF Middle East Command and surprised most observers, showing skill and determination though losing by three goals to five.

At the end of April 1938, the REAF had the following aeroplanes: 20 Avro 626s plus one damaged and currently being used for rigging instruction; three Gipsy Moths, six "Egyptian" Audax VIs; 18 "Egyptian" Audax Xs; nine Magisters; one Avro 651; one Westland Wessex; and one Avro 642 Commodore. Of these the newly arrived Panther X powered Avro 674 "Egyptian" Audaxes were the most potent. Eighteen had been ordered for a proposed new bomber squadron (No. 4) to be delivered between December 1937 and April 1938, after which some of the older Audax VIs would be transferred to the FTS. It clearly took time to get No. 4 (Bomber) Squadron operational, but 17 new machines had arrived by the end of May. No. 4 Squadron then moved to Dakhailah which had officially been its home since late May. Not only did the upgraded "Egyptian" Audax have a better engine but its braking and fuel systems had been improved, resulting in one of the most popular aeroplanes operated by the REAF.

More examples of the Miles M.14 Magister arrived during 1938: L210 (c/n 637) delivered in May, L211 to L214 (c/n 812 to 815) delivered in July, L215 (c/n 816) delivered in August, L216 to 218 (c/n 817 to 819) delivered in September, L219 (c/n 820) delivered in October, and L220 (c/n 798) delivered in November. Unfortunately, the Egyptian FTS soon found itself suffering from the same problems with the Magister as the RAF had experienced. There were three fatal accidents when these aeroplanes failed to pull out of dives which could neither be corrected nor explained. Training with the new aeroplanes was halted and a board of enquiry, on which the Miles Company was represented, found that the control surfaces were inadequate. Modifications were therefore made before the machines were returned to service. This seems to have been associated with the spinning problem seen in Britain. Initial efforts by the Miles Company to cure it by increasing the depth of the rear fuselage, adding strakes ahead of the tailplane and expanding the size and aspect-ratio of the rudder resulted in the Miles M.14A Magister, which was the version purchased for the REAF. However, the spinning problem was only solved with further changes, including raising the height of the rudder. This resulted in the Miles M.14B Magister, which would also be supplied to the REAF.

By late May 1938, six ex-RAF Fairey Gordon Mk. I general-purpose aircraft had been requested for the REAF. They were to form a Target Towing Flight to help train a forthcoming fighter squadron which was to be equipped with Gloster Gladiators. This TT Flight would also help the Egyptian Army's increasing important anti-aircraft batteries. The machines in question were drawn from RAF stocks in the Middle East and some, including number K2705, had previously been used by No. 47 Squadron RAF in the Sudan before that unit was partially re-equipped with Vickers Wellesleys.

The RAF Advisory Mission report of late May stated that the Egyptians were not only ordering 18 Gloster Gladiators (plus a further nine Magisters) but also wanted one or two Hawker Hurricanes so that the REAF could test them, to find which was the best version for Egyptian conditions. Apparently, someone with access to Jane's *All the World's Aircraft* had told the Egyptian Prime Minister that the Hawker Hurricane was faster than the Gladiator, so the British Advisory Mission had to explain that Gladiators could be obtained a year earlier than Hurricanes. Furthermore, the type was well-proven and more suited to inexperienced pilots, had an air-cooled engine and was already in service with the only RAF fighter squadron, No. 33, currently in Egypt. The Prime Minister persisted in wanting Hurricanes but eventually gave way and 18 Gladiators were ordered.

The work of the REAF's FTS in the north-eastern part of Almaza aerodrome really got going in January 1938. Its four

An REAF officer explaining to cadets from the Military College the potential of the REAF's first fighters, perhaps when No. 5 Squadron, was formally declared operational. The Gloster Gladiators were Mk. I machines which had been converted to Mk. II standard. The small oil filters beneath the engine cowlings would mostly be replaced by larger sand filters, as were already fitted to No. 2 Sq's machines. (EAF Museum collection)

Gloster Gladiator Mk. Is of No. 33 Squadron RAF flying over the Suez Canal in Egypt. Number K8053 would be destroyed in an accident on 31 May 1939, though the pilot escaped. Number L7619 later went to No. 112 Squadron but was shot down on 10 November 1940, with the loss of its pilot, Flight Lieutenant K.H. Savage. (Albert Grandolini collection)

to most of the REAF and was also used as a civilian airport. Nevertheless, the REAF's FTS was obliged to remain there for several years.

The main task of the FTS was, of course, to increase the number of Egypt's military pilots, of whom there were only 50 in 1938. In practice the REAF was responsible for initial training, with a one-year course including a minimum 100 hours' flying time. The plan was that Flying Officers would then be sent for further training in Britain or India. By December 1938 the Egyptian FTS had passed out its third flying course or "cohort" and had acquired an American Link Trainer, an early form of static flight simulator which could be enclosed for simulated night or blind flying. Graduates from the FTS at Almaza usually then went to the RAF for further instruction. For example, those sent to the British No. 4 FTS at Abu Suwayr close to the Suez Canal did air interception training over western Sinai, along with their British and British Imperial colleagues. Others were sent to Britain though whether any went to India remains unclear. This further training tended to be intensive and unfortunately resulted in a number of accidents, *Tayyah Thani* Muhammad Mustafa Isma'il being killed in a crash near Frome in England on 9 March 1938 while on a course at the RAF's Central Flying School.

Another, more experienced officer, Hassan Tawfiq, was sent to England to attend a wireless engineering course at Cranwell College. He was there during the Munich Crisis of September 1938 and was still there when the Second World War broke out. At Cranwell, Tawfiq studied alongside some Iraqi officers, one of whom would remain a close friend of the family for many years (see Chapter Four). Tawfiq's daughter, Mona, remembered him well:

British staff (a Chief Instructor and three others), plus six Egyptian instructors (four flying instructors, one navigation instructor and one armaments instructor), had at their disposal four Gipsy Moths, nine Magisters, ten Avro 626s and six "Egyptian" Audaxes. Twenty-two cadets had been accepted from the Army and Police Colleges, but four were found to be unsuitable, probably on medical or aptitude grounds. Eight of the REAF's existing NCO pilots were also accepted on this first course. However, it soon became apparent that Almaza was far from being an ideal location for the FTS, especially when the students reached the solo stage of their flying training. The crowded aerodrome was already home

His name was Yayha Ithniyani. He was a wonderful person and we children called him Uncle Yayha. He was in and out of jail all his life. That was typical of Iraq, where if you joined one group you were considered anti-another group. His wife was Antie

Biplanes of No. 4 Flying Training School during interception or air gunnery training near Abu Suwayr in 1938. (Woodroffe family archive)

Maghlum, who was a great man in the diplomatic corps, an ambassador. We saw her again in Egypt in the 1950s.

In December 1938 an unnamed senior REAF officer undertook a tour of several RAF stations and aircraft manufacturing factories in Britain, with a view to setting up the REAF's own aircraft repair depot in Egypt. Meanwhile the REAF had been tasked with what had become a traditional operation, photographing a 180km stretch of the Nile north of Assyut in May, just before the annual flood, checking for unusual erosion of the riverbanks. The results were then assembled by the REAF's relatively new photographic section at Almaza before being handed to the relevant ministry.

Wahida; she said that Yahya learned to knit in jail and made things for other prisoners. He became very well respected and well known amongst Iraqis of the older generation. But Auntie Wahida got fed up waiting for him being in and out of jail and so got divorced. They had a daughter named Mona after me (Mona Tawfiq) because they were in England together. Wahida did not speak English, getting the words kitchen and chicken muddled up. They all lived at Cranwell together, and mummy helped her with English. After the divorce, Auntie Wahida married Midhat

Unfortunately, May 1938 also saw a report from the British Advisory Mission, stating that Egyptian Army interference in REAF affairs had resulted in its men and machines being used for political purposes such as flying displays to impress the populace and escorting members of the royal family or government officials. On the other hand, an air firing and bombing exercise on 1 May 1938 proved a success while a similar display in front of King Faruq on 1 November demonstrated an accuracy well above average.

The next intakes of cadets into the Military College in 1938 included some who would become politicised, but more who would rise to prominence because of their military activities. Those who eventually joined the REAF included Muhammad Abu Zaid and Adli Kafafi. The latter was born on 21 March 1921, gained his secondary school baccalaureate at the age of 15 but had to wait until he was 16 before joining the military. In the meantime, he studied medicine, in case his dream of being an airman came

The REAF's training schedule was often interrupted by the need to put on a display for some important national event in Egypt. Other air forces had to do the same, but the REAF was still so small that the same aircraft and crews tended to be called upon. Here three flight of Avro 626s and one of Miles Magisters pass over the Qubbah Palace in Cairo, during celebrations to mark the marriage of King Faruq and Queen Farida in 1938. (Author's collection)

Hassan Tawfiq and an Egyptian colleague with an Armstrong Whitworth Atlas TM Trainer while training in the RAF's No. 4 FTS at Abu Suwayr around 1938. (Mona Tewfik collection)

Hawker Hart Trainers over RAF College Cranwell in the mid-1930s. A small number of those regarded as the best in the Royal Egyptian and Royal Iraqi Air Forces were selected for further training at this highly prestigious college in Lincolnshire, England. They included Hassan Tawfiq who attended a wireless engineering course and was there during the Munich Crisis of September 1938. (Fleetway archive)

The oasis town of Siwa with its hilltop castle, seen from the air around 1938. Siwa was a vital Egyptian military outpost a few kilometres from the frontier with Italian-ruled Libya. (EAF Museum collection)

to nothing. After that Adli Kafafi entered the Military Academy for one year, probably starting in November 1938. Adli Kafafi's parents, especially his mother, were not keen on their son giving up medical studies to be an officer but his mother eventually accepted that Adli's passion for aviation was more than merely a hobby. The son of Adli's school principal, Muhammad Abd al-Rahman Kassab similarly joined the Air Force.

In her book *In Memorian 15 May 1948: When my father fell, a tune upon the strings of my memory*, Munira Kafafi described how she spent a long time talking to General Umar Tantawi Pasha, a teacher at the Egyptian Army Military College during the period when it was led by *Amir Alay* Muhammad Fattuh. This was after the retirement of a British officer named Thorburn following the Anglo-Egyptian Treaty of 1936. Amongst Adli's tutors was Muhammad Mitwalli who would subsequently command the REAF from November 1944 to early 1948. General Tantawi also described the entry of 1938 as "a merry crew" who happily studied military science, engineering, Arabic and the English language.

In November the following year, Adli Kafafi went to the Flying Training School where he gained his pilot's wings in 1940. Nevertheless, the British Military Mission's report of November 1939 mentioned continuing dissatisfaction amongst REAF cadets at being put on a separate seniority list from that of the Army. Furthermore, Adli's daughter Munira maintained that her father was associated with Gamal Abd al-Nasser's nascent Free Officers Movement, though he was killed during the Palestine War before coming to political prominence.

While the training of Egyptian aircrew received most publicity, the training of ground crew was equally important. The Air Force's first such technicians and mechanics had been recruited amongst existing, partially trained men from the country's limited industrial base, plus assorted craftsmen. By now, however, their recruitment and training were on a more regular basis. For example, on 23 March 1938 a group of students from the Muhammad Ali Industrial School visited the RAF's Depot at Abu Qir on the north-western coast of the Nile Delta, and during September that year the REAF's own Mechanics School began its first training course.

Then came the Munich Crisis of September 1938. At this time the main aerial threat to Egypt was from the Italian *Regia Aeronautica* and in October, in the aftermath of the resulting Munich Agreement, Mussolini tried to use the widespread atmosphere of relief that war had been avoided to press his own demands upon the British and French. Amongst other demands, the Italian dictator wanted Italian participation in the running of the Suez Canal Company, which was still owned by the Compagnie Universelle du Canal Maritime de Suez. Italy was especially opposed to the largely French company's monopoly over the Suez Canal because all Italian merchant traffic to its East African colonies was obliged to pay canal tolls. Mussolini also expected that, because Italy had helped defuse the Munich Crisis, Britain would be favourable to this point of view and thus pressure France to yield to Italy's demands. In fact, neither France nor Britain was willing to go along with such ideas, which in turn encouraged Hitler to declare that Germany would come to Italy's support in case of conflict.

For the REAF, the Munich Crisis and its aftermath meant that a small number of "Egyptian Audaxes" and Avro 626s were hurried to Marsa Matruh and Siwa Oasis, to again patrol the Libyan frontier and neighbouring Egyptian coastline. Meanwhile the Egyptian government declared the "Western Oases" of Bahariya, Dakhla, Farafra and Kharga along with Siwa to be prohibited areas. The Italian invasion of Albania, plus Nazi Germany's seizure of what remained of Czechoslovakia appalled almost everyone in Egypt, including King Faruq. Even the previously popular Marshal Italo Balbo found himself received coldly by the Royal Palace in May 1939, after which Italian propaganda took a less sympathetic view of Egyptian aspirations.

The REAF's first modern military aircraft were Westland Lysanders, here lined up at Almaza shortly after their delivery to No. 1 Squadron in November 1938. Most would be repainted in a camouflage scheme early in the Second World War and a handful of survivors were still in service during the Palestine War or 1948. (Author's collection)

One of the REAF's new Westland Lysanders (number Y517) before it was delivered to Egypt late in 1938. (Author's collection)

In September, No. 208 Squadron RAF was similarly rushed to Marsa Matruh from their training at Abu Qir. This was apparently done at one day's notice, so that the squadron could cooperate with the Egyptian Army's Mobile Division. In fact, the British may have reached the frontier zone before the REAF's machines. Because of a lack of facilities at Marsa Matruh, all aeroplanes still had to operate from the beach while those at Siwa were similarly exposed at a rudimentary desert airstrip. From these outposts they patrolled the Libyan frontier, occasionally meeting Italian Savoia-Marchetti tri-motors doing the same thing on the other side of the 320km-long barbed wire fence that the Italians had erected from Jaghbub to the coast (see Volume 3). As before, three Egyptian aircraft would fly each patrol which now consisted of one Avro equipped with a radio, and two Audaxes, each with a single Lewis gun. These were still the only Egyptian aeroplanes to carry weapons, though they could not be used because some vital parts had yet to be delivered. Furthermore, the crews still had virtually no weapons training.

Nevertheless, three aircraft flying at a steady 120 metres altitude would follow the coast from Marsa Matruh to the frontier and thence down the frontier to a landing ground on the plateau 19km beyond Siwa Oasis. The Flight Commander's aircraft was still identified by a pennant with a small Egyptian flag attached to the interplane struts, in a system dating back to the First World War. Among those who led such patrols was *Qa'id Sirb* Ibrahim Hassan Gazarine who subsequently rose to the rank of *Qa'id Firqah Jawwiyah* (Air Commodore) in the REAF.

During this period, the Hawker Hinds of recently re-established No. 113 Squadron RAF were similarly based at Marsa Matruh, making a photographic survey of the frontier zone. In November plans for a proper aerodrome at Marsa Matruh were reportedly complete, though no work had started in the buildings. The REAF would continue to fly such frontier patrols until Italy entered the Second World War in 1940. Meanwhile most of the REAF's Army Cooperation squadrons, which had cooperated with the British Army (Suez) Canal Brigade in August 1938, remained on standby throughout the September emergency before returning to their usual training duties. The squadron was meanwhile due to exchange its Avro 626s and "Egyptian Audaxes" for new Westland Lysanders.

The scare caused by the Munich Crisis gave added urgency to what was already a change in official British attitudes towards Egypt and its armed forces. The REAF would now receive modern

– though not the most modern – equipment, including two squadrons of Gloster Gladiator fighters, one of Westland Lysander army cooperation aircraft and one of Avro Anson bomber-reconnaissance machines. The Westland Lysanders arrived first and were the first modern aircraft to be flown by the REAF, having been ordered for No. 1 Squadron back in 1938. Perhaps more significantly, the British agreed to supply Egypt with adequate spares for its new machines, urging upon the Egyptian authorities what some Egyptian historians have described as an emergency expansion programme during the period of panic between the Munich Crisis and the outbreak of the Second World War.

Not that the Egyptian government was reluctant to expand its armed forces – far from it. However, those who actually ran the REAF, Egyptians and British advisors alike, knew that the tiny, understaffed and inexperienced Egyptian Air Force was barely in a position to absorb much new hardware. It simply did not have the infrastructure and personnel required. Time also seemed to be needed for some deeply engrained attitudes and traditions to change. For example, it was only in November 1938 that the Reports of the British Advisory Mission started to refer to the Egyptian Air Force as the REAF rather than the EAAF. On a more symbolic level, the REAF adopted light blue RAF style uniforms, though these seem to have been of a darker shade than their British counterparts, while the red *tarbush* or fez continued to be worn with ceremonial dress for several years. The Egyptian Army and the Military College continued to wear it for considerably longer, not only with dress uniforms but as part of their everyday service wear.

According to a British report of 31 October 1938, the REAF now had 71 aeroplanes: 18 "Egyptian Audax" Mk. Xs; six "Egyptian Audax" Mk. VIs; 21 Avro 626s; four Gipsy Moths; 19 Magisters; one Avro 652; one Wessex; and one Avro Commodore. Nevertheless, the force continued to face technical problems which limited its operational capability. Thus, the British Advisory Mission report for the period 1 May to 31 October 1938 noted that No. 4 (Bomber) Squadron at Dakhailah previously had trouble with its petrol tanks during the hot summer weather, resulting in parts of the Egyptian press criticising the reliability of British aeroplanes. A lack of ground facilities at Dakhailah held up its training but problems with overheating Audax engines had been solved. Similarly, the six Fairey Gordons acquired as a Target Towing Flight were still denied towing gear.

Recent reorganisation also caused confusion in the maintenance schedule. The Egyptian Ministry of War remained short-staffed and overburdened while the Air HQ at the Ministry was described as "entirely inadequate". In December Egypt's Supreme Defence Council agreed a further reorganisation and rearmament of the REAF, with over a tenth of the annual defence budget being set aside for an ambitious five-year plan. This envisaged a substantial number of military aerodromes being constructed across the country and aimed for an Air Force of no less than 500 new aeroplanes, along with a broader five-year plan for the Army and a revived Navy. The overall defence budget was set at 40 million Egyptian pounds, a massive sum which many thought unrealistic, but War Minister Hassan Sabri Pasha threatened to resign if he did not get it. The money was duly allocated and Hassan Sabri remained in post.

As of 11 November 1938, the REAF had 49 Egyptian officers, 1,291 Egyptian other ranks and 24 British advisors. No. 1 General Purpose Squadron under *Qa'id Sirb* Salih Mahmud Salih was eagerly awaiting new Mark I Westland Lysanders which would result in it being redesignated an Army Cooperation squadron. Eighteen had been purchased and would be given serial numbers Y500 to Y517 and, together with an ex-RAF Mk. I (RAF serial number R2650) and an ex-RAF Mk. III (RAF serial number R9000) that were handed over during the Second World War, would soldier on through assorted military and political crises until 1950. For its part, the REAF's FTS awaited the arrival of a substantial order of new Miles M.14B Magisters.

The latter half of 1938 saw the arrival of some of the REAF's new aeroplanes, including three Avro Anson Mk. Is which had been diverted from an order intended for the RAF. They would form the backbone of a Bomber/Transport squadron and although they initially went to No. 3 (Communication) Squadron, the Ansons joined No. 4 (General Purpose) Squadron later in 1939. They were given REAF serial numbers W205 to W207, having previously been allocated RAF numbers N9648, N9657 and N9666.

Once redesignated as an army cooperation unit, No. 1 Squadron received its first two Lysanders in late November 1938, the rest of this order arriving by the end of February 1939. No. 1 Squadron's aircrew could now start training on their new aeroplanes at Almaza alongside No. 208 Squadron RAF. According to British Advisory Mission records, the Egyptians were taught first and then helped instruct their British colleagues who were themselves the first RAF unit in the Middle East to receive Lysanders.

The first batch of 18 Gladiators was slower to appear and even at the end of 1938 the Fairey Gordons of the Target Towing Flight still had no towing gear. Meanwhile the Egyptian government investigated the possibility of acquiring Bristol Blenheim bombers for No. 4 Squadron, currently based at Dakhailah, the idea being for No. 4's "Egyptian Audax" Xs to be transferred to the FTS as advanced trainers. There was also talk of No. 3 Squadron REAF taking over three aged ex-RAF Vickers Valentia transports, though this idea was eventually dropped because of the age of the aeroplanes and especially their engines.

Prince Abbas Halim meanwhile continued to show a keen interest in aviation in general and the REAF in particular. According to his daughter, Princess Nevine Abbas Halim:

> In September 1938, our last summer in England before the war, Daddy was invited to Germany.... Mum says he was asked by Goering to make an inspection of the factories in Germany to note what Germany had in the way of airplanes. Abbas brought back a report which he later discussed with the people at the English Speaking Union in London, when we had lunch one day. He told them, what Lindbergh had previously said, that this was a very dangerous business [namely that Nazi Germany had built up a potentially very threatening air arm]. But people laughed at him and called him a defeatist. He repeatedly said; 'I'm telling you what I've seen with my own eyes.' In Cairo, he told the British about his factory visits in Germany. They would not believe him either.

In his next speech from the throne on 19 November 1938, King Faruq announced that his government would establish armaments factories – something that had not been seen in Egypt since the reign of Khedive Muhammad Ali in the first half of the 19th century. The military service law would also be amended and substantially more officers trained. Two days earlier King Faruq and Queen Farida had their first child, Princess Farial, followed by celebrations across Egypt, including the distribution of clothes

The Aerogypt "safety aeroplane" was designed and built by Salih Hilmi (Saleh Helmy), an Egyptian engineer. With the British civil registration G-AFFG, the Aerogypt Mk. I first flew at Heston aerodrome in west London in 1938. It incorporated an additional aerofoil section on top of the fuselage which could be raised as a landing flap and also had a slow speed automatic slot at the front. (Author's collection)

aircraft units being formed in 1938. It was hoped that they would remain under some degree of British control, to be integrated into the broader British defence plan for Egypt. This also involved a highly secret British "remove radar station" sited close to the main road between Cairo and Suez. It was reportedly in operation by November 1938.

Quite how keen the British were on Egypt regaining a real Navy, as distinct from an armed Coastguard, is open to question. Nevertheless, in February 1938 Egyptian personnel started training aboard four British Royal Navy minesweepers. By April they were operating the two minesweepers which Britain agreed to lend to Egypt and which would also be based in Alexandria. Meanwhile the Egyptian Ministry of Defence began recruiting 100 naval personnel.

and free breakfasts to the poor. The families of the estimated 1,700 Egyptian children born the same day were given one Egyptian pound, a substantial gift at that time, especially amongst the country's poor.

One pilot of the RAF's No. 113 Squadron, whose Hawker Hinds were still doing a photographic survey of the frontier with Libya, recalled that he and his colleagues now shared the only hotel in Marsa Matruh with the men of a Flight of the REAF's No. 4 Squadron. A few days after the birth of Princess Farial, King Faruq and Queen Farida suddenly turned up to celebrate becoming parents. Apparently, they had flown at night from Cairo in 'an early form of Anson', in other words the Avro 652, whose crew had supposedly 'never flown at night before'. Another story from a different source claimed that only a few months earlier, in August 1938, the CO of No. 3 Squadron gave the young King Faruq his first experience of flying with the REAF, probably in the Avro 652. Faruq was clearly impressed as he promptly declared that he must have his own Royal Flight.

While the King and his government were planning to establish armaments factories in Egypt, other Egyptians were venturing into the design and building of aeroplanes. One was Mr Salih Hilmi (Saleh Helmy) whose unusual four-seat monoplane incorporated an aerofoil into the shape of its fuselage to provide additional lift. It was powered by three Douglas Sprite engines, based upon the British made Douglas F/G31 motorcycle engine. Furthermore, this Aerogypt I, or "Safety Plane" as it was called, had a hinged cabin roof which, when raised, acted as a landing flap. It first flew at Heston aerodrome in west London in 1938 with the British civilian registration G-AFFG and, though it was not particularly successful, Mr Hilmi's Aerogypt explored an aspect of aircraft design which continues to fascinate aero-engineers.

Following the Anglo-Egyptian Treaty of 1936, the British seemed keener on providing the Egyptian Army with anti-aircraft defences than they were to strengthen the REAF, the first such anti-

The seriousness with which Britain and Egypt viewed the looming threat of war was reflected in a new agreement concerning the anticipated role of the Egyptian armed forces. It is clear that the governments of both Egypt and the United Kingdom expected Italy to enter any conflict as an ally of Germany at an early date, as reflected in the plan outlined in a secret letter sent by H.R. Nicholl at HQ RAF Middle East to Husain Sirry Pasha, the Egyptian Minister of Defence, on 23 January 1939:

War Role of the REAF Army Co-operation Squadron.

(a)   It was agreed that one Flight of this Squadron would, in war, be employed on duties in connection with the defence of the Suez Canal, and that its main role would be reconnaissance of the Gulf of Suez and the Northern part of the Red Sea.

It was agreed that Suez would be the most suitable base for the Flight, and that the Director, REAF, would arrange for the accommodation of the Unit at Suez for the communications between the Unit's base and the Headquarters of the [British] Officer Commanding, Canal Defence Forces.

The advantage of having this unit located at Suez rather than at Ismailia is that it shortens the length of operational flight required for reconnaissance in the Gulf of Suez and, moreover, the presence of Lysander aircraft there would act as a partial deterrent to enemy air attack on Suez in that this modern type of aircraft possesses characteristics which qualify it to undertake work in defensive air fighting [a peculiarly optimistic view of the Lysander's capabilities]. In the event of an emergency arising before arrangements at Suez have been completed, the Flight will be based at Ismailia as was done in the case of the Squadron in the recent emergency [the Munich Crisis].

(b)   It was agreed that the second Flight of the REAF Army Co-operation Squadron should be employed in reconnaissance and bombing duties in the Western Desert in the Bahariya area. The primary role of the Flight would be to locate enemy movements from the frontier [with Italian-ruled Libya].

The Director of the Royal Egyptian Air Force undertook to ensure that the aerodrome at Bahariya [oasis] was kept in readiness for the occupation by this Unit, and that the necessary supplies of fuel, bombs, etc. were maintained there. It is to be understood, however, that the Unit will not actually move to Bahariya until the protection of this place against land attack has been ensured by the Army.

If protection against land attack cannot be guaranteed, it is proposed that the Unit should remain based at Almaza, but should make use of Bahariya as an advanced operational landing ground for refuelling, etc.

War Role of the REAF Fighter Squadrons.
It was agreed to be desirable that the Fighter Squadrons of the Royal Egyptian Air Force should be allocated in the first place to the "home defence" of Cairo and Alexandria areas, and that No. 1 (Fighter) Squadron, when formed, should be allocated to the defence of Cairo.

It was agreed that this Squadron should in war be located at Helwan, which will be the peace time station of the Royal Air Force Fighter Squadron which, in war, moves to the Western Desert. (One Fighter Squadron of the Royal Egyptian Air Force will also be located at Amiriya for the defence of Alexandria and the Fleet Base).

It was also agreed that the operational control of all Fighter Squadrons employed in "home defence" should, in war, be under one command, and that this Commander should be stationed at the "Air Raid Report Centre" now being established at Dekheila Aerodrome. Until the time when sufficient Egyptian Fighter Squadrons are available to take over this duty it is proposed that a British Air Force officer should, in war, direct the operations of all the Fighter Squadrons employed in "home defence", and that with him should be an Egyptian Air Force officer who would issue the required orders to the Egyptian Air Force Fighter Squadron.

With regard to the Air Raid reporting system and Report Centre which is at present being organized, it is proposed to hold in the near future an Exercise in the Alexandria area in which the Royal Egyptian Air Force and the Royal Air Force Fighter Squadrons will co-operate from their War Stations.

The formation of the second Royal Egyptian Air Force Fighter Squadron was discussed, and it was agreed that it would be an advantage if this could be expedited, as it is considered that the presence of REAF Fighter aircraft operating in the defence of Cairo and Alexandria would have a most beneficial moral effect on the population, in addition to their material effect as a deterrent to enemy air attacks on these important centres. It was therefore agreed that the provision of further Fighter Units for the Royal Egyptian Air Force should be given priority over the provision of Bomber [transport] or other type Units.

War Role of the REAF Bomber Squadron.
It was agreed that until such time as the second Fighter Squadron of the REAF has been formed and is available to take its place in the defence of Alexandria, the flight of Panther Audax Bomber aircraft should, in war, remain at Dekheila to act in the first place as a reserve for short distance reconnaissance or offensive action against enemy forces in the Western Desert.

Preparation of Operational Landing Grounds in the Western Desert.
It was agreed that the Director of the Royal Egyptian Air Force should keep me informed of the progress made in this important work of preparing operational landing grounds in the Western Desert, and that I should keep him informed of the requirements of the aerodromes for operations.

Corridors and Prohibited Areas.
The necessity for providing corridors of approach to defended areas for Egyptian and British aircraft and for laying down prohibited areas was discussed, and it was agreed that details of the arrangements proposed should be supplied to the Director of the Royal Egyptian Air Force by me, but that, in order to ensure secrecy, they will not be issued down to Units of either service until the time arrives when it becomes necessary to put them into force.

Employment of Misr Airways in War.
It was agreed that in the event of war, it would be of great value if Misr Airways could be taken over complete by the Director, Royal Egyptian Air Force, and that the personnel and aircraft employed for the operation of a transport service for the evacuation of casualties or other work. If this scheme is approved, it is proposed that the Air Officer Commanding, Royal Air Force, shall approach the Air Ministry for approval for British personnel employed with Misr Airways to continue to operate with them in war.

Air Raids Reporting System.
With regard to the Air Raids Reporting System, the Director, Royal Egyptian Air Force is at present engaged in organizing a Report Centre at Dekheila. This scheme, as at present organized, will only partially meet all requirements for obtaining reports of enemy aircraft about to attack, and warning military units and civil population, and the Air Officer Commanding, Royal Air Force is at present engaged, in collaboration with the General Officer Commanding-in-Chief, in drawing up a comprehensive scheme for Air Raid Reports and warnings which it is hoped will be communicated to Your Excellency [Husain Sirry Pasha, Egyptian Minister of Defence] for approval in the near future.

I shall be grateful if Your Excellency will approve in principle the foregoing arrangements which have been agreed with the Director, Royal Egyptian Air Force, with regard to the role of the REAF in war and the other subjects mentioned, and if you would approve the continuance of the system of direct communication and discussion of details between the Director and myself.

This plan was agreed by Husain Sirry Pasha on 12 February 1939. However, it is worth noting that according to some sources there were not as yet any Egyptians amongst the "first pilots" of Misr Airway's multi-engine aircrew, the most experienced still usually flying as co-pilots or "second pilots". The expansion of Misr Airways, or Misr Airlines as it was also called, had been rapid and ambitious both in terms of aeroplanes and routes flown. Despite the supposed – and in practice mythical – pro-German bias of some senior members of Misr Airways personnel, the British De Havilland company remained central to this programme.

Operations had begun back in July 1933, linking Cairo with Alexandria and Marsa Matruh using De Havilland DH 84

Dragons. On 3 August 1935 Misr Airways started a test service to Nicosia in Cyprus via Lydda in Palestine using De Havilland DH 86 Express aircraft but this was ended on 20 October after less than three months.

The four-engine DH 86 was an enlarged, more streamlined version of De Havilland's highly successful DH 84 Dragon, powered by the most powerful 200 hp Gipsy Six engines currently made by De Havilland, the main production version also having two pilot's seats, side by side. When the first production aircraft entered service in October 1934 it was the fastest British-built passenger aircraft, and the subsequent dual-pilot type with its lengthened nose unexpectedly proved to be even faster. Unfortunately, the DH 86 Express suffered from a serious lack of directional stability, resulting in a series of fatal crashes. Investigations in 1936 resulted in late production machines being given additional vertical so-called "Zulu Shield" extensions to the tail planes to increase the fin area, the resulting aircraft being designated the DH 86B.

By April 1939 the Misr Airlines fleet consisted of one De Havilland DH 84 Dragon (SU-ABI, c/n 6031 which crashed at al-Arish on 15 March 1935), two De Havilland DH 86 Express (SU-ABN c/n 2320 and SU-ABO c/n 2329), one De Havilland DH 86B Express (probably SU-ABV [ex G-AJNB] c/n 2342 acquired on 1 January 1937, though this is described elsewhere as a DH 86A), five De Havilland DH 89 Rapides (SU-ABP c/n 6298, SU-ABQ c/n 6299 destroyed in an accident on 9 October 1941; SU-ABR c/n 6302; SU-ABS c/n 6303; SU-ABU c/n 6313) and one De Havilland DH 90 Dragonfly (SU-ABW c.7553) acquired on 22 July 1937.

The Misr Airlines De Havillands now operated the following routes: Alexandria – Cairo; Alexandria – Port Said – Cairo – Miniya – Assyut; Cairo – Assyut – Luxor – Aswan; Cairo – Lydda – Haifa – Baghdad; and Cairo – Port Said – Lydda – Haifa. Control of these routes was taken over by the Egyptian government in September 1939, following the outbreak of the Second World War, and in the following year, 1940, another service to Beirut and Palestine began. (By this time, according to his son Usama, Muhammad Sidqi's pet crocodile – which had accompanied him during his flight from Berlin to Cairo in 1930 – grew too big to be kept at home as a pet. It was therefore given to Cairo Zoo, where it lived for many decades – apparently still being there when the author visited Cairo Zoo in 1963, and even when the author interviewed Usama Sidqi 35 years later.)

The targets in Egypt which were most vulnerable to air attack were considered to be the Alexandria Naval Base, the Port Said oil refinery and oil storage tanks, the facilities of the Suez Canal Company, the forward military base at Marsa Matruh, and Cairo itself in an effort to undermine support for the Egyptian government. At this time the British feared an Italian airborne rather than overland assault to cut the Nile Valley somewhere between Aswan and the Sudanese border at Wadi Halfa, launched either from al-Uwaynat in the far south-eastern corner of Libya, or more likely from the larger oasis complex of al-Kufra. The Italian Army was, in fact, training a substantial unit of indigenous Libyan paratroopers who might have been particularly useful in such an assault (see Chapter 2). In the event, the Italian Army sent just a few reconnaissance patrols from al-Uwaynat across the frontier into Egyptian territory during 1940 before withdrawing from this lonely outpost by January 1941.

A Spartan Cruiser II (registration SU-ABL) of Egypt's Misr Airways, in company with various British aeroplanes at Heston airport, Middlesex, in the mid-1930s. Built in 1933 and initially registered as G-ACDW, this Spartan Cruiser II was sold to Egypt in March 1934. (Author's collection)

The De Havilland DH 84 Dragon was a successful short-haul airliner that was sold to many countries. Rugged, efficient and relatively simple to maintain, three were purchased in 1933 by Egypt's airline Misr Air, whose SU-ABJ is seen in flight above the desert. They would be followed by the more advanced Express and Dragon Rapide, and would be called upon to support the Allied war effort a few years later. (Ahmad Isma'il collection)

The way in which much of Egypt's educated elite embraced "western ways" during the 1930s is illustrated by this photograph of a Misr Air trainee air hostess with one of the airline's new De Havilland DH 86 Express aeroplanes. Egypt bought three (SU-ABN, SU-ABO shown here and SU-ABV) in 1935 and 1937, of which the third was a DH 86A. (Ahmad Isma'il collection)

In August 1939 Egyptian and British troops therefore moved to their agreed positions. For the Egyptians the most active of these were two squadrons of the Frontier Force based at Siwa Oasis and at Sollum, at the northern end of the Egyptian frontier with Libya. From these locations Egyptian Army patrols could report any hostile movements. It was an arrangement designed to avoid "provoking" the Italians by placing too many British next to the Libyan frontier while at the same time hampering any Italian attempt to use the desert route south of the Qattara Depression to smuggle weapons to the estimated 50,000 supposedly "pro-fascist" Italians currently living in Egypt.

There were therefore several important reasons why, in March 1939, the British War Office sent urgent supplies of 2-pdr anti-tank guns, Bren guns, Boys anti-tank rifles, ammunition for mortars and wireless telegraphy equipment to the Egyptian Army. Some would still be in use during the Palestine War, almost ten years later. Yet there was still doubt about the reliability of Egyptian troops, a subject which came up in a conversation between Mr Smart, the Oriental Secretary at the British Foreign Office, and Prince Muhammad Ali Tawfiq who had been the chief regent prior to King Faruq's coronation. He was also the heir presumptive to the Egyptian throne from 1936 until the birth of King Faruq's son in 1952. The prince was worried about the impact of Italian and German propaganda in the event of war, assuring the Oriental Secretary that Egyptian troops would not defect to the enemy but fearing that they would not fight with enthusiasm. The British responded by suggesting that Libyan refugees in Egypt be given Egyptian nationality and be sent to foment discontent in Libya. In April, the British Foreign Office discussed a similar idea that Egyptian Muslim agents be used to foment unrest in Italian Libya, this having been proposed by Abd al-Rahman Azzam, the Egyptian ambassador to Iraq, Persia and Saudi Arabia.

Despite reservations about the enthusiasm of some in the Egyptian Army, August 1939 saw its proposed role in case of war being agreed between the Egyptian Minister of War and Lieutenant General Henry Maitland "Jumbo" Wilson who was now IN command of British forces in Egypt. Though essentially the same as the proposal drawn up earlier in the year, it was not yet a finalised plan. Units of the Egyptian Army would protect the railway between Alexandria and Marsa Matruh, contribute troops to defend Alexandria, provide nine battalions for anti-sabotage duties, as well as anti-aircraft units at both the northern and southern ends of the Suez Canal, plus other coastal defence units. This would strain the small Egyptian Army to its limits but was considered vital in order to free British troops for offensive operations.

From January 1939 to the outbreak of the Second World War at the start of September, the REAF worked hard to expand, train and carry out its increasingly serious war preparations. One problem remained – finding enough medically suitable recruits amongst the numerous volunteers. In fact, the British Advisory Mission thought that the Egyptian medical examination board was probably too zealous, excluding men who would have passed as A1 fit in other countries. There was also a serious failure rate amongst those selected for flying training and at the end of one course only four of the six officer cadets had passed while none of the 11 NCO cadets did so, largely because of a lack of basic education.

During the winter of 1938–1939 the REAF's FTS passed out its Third Course or "cohort" and it was at this time that a young RAF NCO named Frederick Weston arrived at Almaza. He formed part of the British Advisory Mission and served as the engineering advisor at Almaza, working with the Flying Training School. Weston recalled that the REAF flying cadets – of whom there were about 200 during his prolonged time at Almaza – were usually between 18 and 19 years old, mostly from prosperous or aristocratic families. He gave lectures on the theory of flight, and the design and construction of aeroplanes and their engines. In return these keen youngsters taught him Arabic and as a result Fred Weston got to know some of them well.

It is worth noting that several of the more senior front-line REAF pilots during the 1948–1949 Palestine War had been commissioned in 1939 and or started to learn to fly in the FTS's Fourth Course. The most detailed information focuses upon those who were killed during the Palestine War, when their obituaries were published in the Egyptian press. They included Muhammad Adli Kafafi, Mustafa Sabri Abd al-Hamid Hasani, Sa'id Afifi Muhammad al-Janzuri, Muhammad Abd al-Hamid Abu Zaid and Najib Abd al-Aziz Basiyuni. Adli Kafafi was commissioned before the Second World War started and was awarded his wings later in 1939. According to his daughter Munira, his senior instructor was Pilot Officer Webb with whom he flew in the Miles Magister and Avro 626. Kafafi then moved to the "Egyptian Audax" on which he was instructed by Egyptian pilots, Mahmud Sidqi Mahmud, Salah Zaki, Muhammad Hassan al-Maghribi, who himself became a flying instructor early in 1939, Muhammad Hafiz and Mahmud Shafiq Habib under the leadership of *Qa'id Sirb* (Flight Lieutenant) Nu'man Nada.

Muhammad Abd al-Hamid Abu Zaid was commissioned as an officer on 1 February 1939, whereupon he was sent from the Military College to the FTS at Almaza and proudly received his new identity card. In this, his next-of-kin was specified as his father Sa'ad Abu Zaid, who was currently

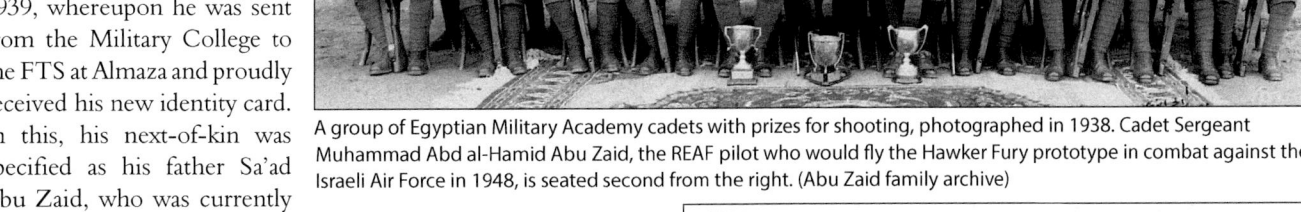

A group of Egyptian Military Academy cadets with prizes for shooting, photographed in 1938. Cadet Sergeant Muhammad Abd al-Hamid Abu Zaid, the REAF pilot who would fly the Hawker Fury prototype in combat against the Israeli Air Force in 1948, is seated second from the right. (Abu Zaid family archive)

aboard the *Star of Suez* of the Alexandria Navigation Company's Red Rose Line. Muhammad Abd al-Hamid Abu Zaid was then photographed with his father, probably in Suez, and wearing their respective Coast Guard and Air Force uniforms.

Muhammad Abu Zaid did not have to wait long for his first flight at Almaza, which was in Miles Magister serial number L218 on 22 February 1939 with *Tayar Thani* al-Maghribi. Abu Zaid's first experience in an "Egyptian Audax" came on 2 March with *Tayyar Thani* Farid when they flew from Almaza to Alexandria in K509. He returned the following day with al-Maghribi in K515. Muhammad Abu Zaid flew a test for his first solo with Nu'man Nada (still ranked as a *Tayyar Awal* in Abu Zayd's logbook) on 25 March 1939 in Magister L226. This was followed by a further test with the British instructor D.R. Britton in Magister L211 on 29 March. This was probably Wing Commander E.A.C. Britton, who was attached to the REAF.

Abu Zaid flew his first solo the following day in Magister L209. During April and May his training largely consisted of flights in a Magister with al-Maghribi, solos in the same type, with one cross country flight with a different British instructor, Warrant Officer Norris, flying from Khanka to Almaza in Magister L242 on the morning of 16 May. However, from 19 May to 2 June Muhammad Abu Zaid did not fly, perhaps being on leave.

From 3 June onwards the same pattern continued, usually with al-Maghribi who was now a *Tayyar Awal*; probably reflecting the fact that one of the REAF's two annual promotions had intervened. Presumably Nu'man Nada had similarly been promoted from *Tayyar Awal* to *Qa'id Sirb*, as shown in Adli Kafafi's records. The end of this page of Muhammad Abu Zaid's log was signed by Abd al-Hamid al-Dighaydi who would shortly be replaced as Officer Commanding Almaza by Nu'man Nada. Here Abu Zaid's proficiency as a pilot was described as "above average" on the Magister. This comment continued: 'He has the makings of a good pilot but to be checked continuously for being overconfident'. Muhammad Abu Zaid's confidence would be noted throughout his career and may have been the cause of his death in 1948.

From 3 to 13 July, Abu Zaid continued to fly Magisters solo or with British Warrant Officer Falconer. On 15 July he had first flights in an Avro 626 (number J326), with Warrant Officer Falconer and Warrant Officer Humphrey, followed by his first solo on the

Muhammad Abd al-Hamid Abu Zaid (far left, second row from back) and other cadets on their graduation from the Egyptian Army's Military College in Cairo in 1939. (Abu Zaid family archive)

type (in J3267) on 20 July. Nada rather than Dighaydi now signed each page of Muhammad Abu Zaid's logbook, countersigned by Falconer. This pattern of training continued throughout August, mostly on the Avro 626 when he flew solo or with Falconer or

Humphrey, plus occasional flights in a Magister. Abu Zaid was still under training at Almaza when the Second World War broke out. In fact, it was on 2 September 1939, the day after Germany invaded Poland and the day before Britain and France declared war, that Muhammad Abu Zaid had his first flight in an "Egyptian Audax" (number K517) with Warrant Officer Falconer.

Ali Sabri was another of those training on the Avro 626 in the REAF's FTS in 1939, as was his elder brother Husayn Zulfiqar Sabri, alongside Abd al-Latif Baghdadi. One of the latter's surviving letters was dated 16 February 1939. It was addressed to his father in the little village of Shawa and explained how the young man had achieved his dream by becoming an Air Force officer and accepted for flying training at Almaza:

My dear Father,
   Greetings and respect,
   I have received your honourable letter and I am so grateful to you for this kindness. I will return this gift one day, God willing, but I would prefer that you do not send me any more money, because I want to be a man who relies upon himself and to win my bread from the sweat of my own brow. I have spent a lot [of your money] for about twelve years but now that I am a man and have achieved what I have wished for a long time, I want to rely on myself in my new life, not on anyone else.

In addition to the FTS, Almaza remained home to several of the REAF's operational squadrons, including those which sent one or more Flights to other parts of the country. Amongst them were the Lysander crews of No. 1 Squadron who had completed training alongside the equally new Lysanders of No. 208 Squadron RAF. In fact, one of the latter's pilots would return to work alongside the REAF in the British Advisory Mission. In January 1939, after conversion to Lysanders, No. 208 Squadron RAF went to its new home at al-Qasaba east of Marsa Matruh. Perhaps typically, No. 208 Squadron's history made no mention of this unit's time converting to Westland Lysanders alongside the REAF at Almaza. Meanwhile No. 1 Squadron REAF continued training, its personnel being photographed lined up for inspection by the REAF's Director, the white-haired *Amir Alay* Ali Islam Bey. Whereas the British Advisory Mission described the latter as "pleasant but totally ineffective", the Mission described No. 1 Squadron CO, *Qa'id Sirb* (acting *Qa'id Asrab*, Flight Lieutenant acting Squadron Leader) Salih Mahmud Salih, as "an outstanding officer".

The British Advisory Mission of which Fred Weston formed a part, was eventually headed by an RAF Group Captain with a Wing Commander as Senior Advisor, plus 12 RAF flying instructors and four RAF engineers – one for each training unit and one for the workshop. Six more RAF flying instructors and two ground instructors had been seconded to the FTS around May. Unfortunately, an Advisory Mission report of June 1939 recorded that the Egyptian authorities refused to accept three of the British instructors recommended by the Mission. Fred Weston himself had been selected late in 1938 as a specialist aircraft engineer. His was to be a three-year contract which involved good pay, the same privileges as the diplomatic corps, and more freedom of action than Fred Weston had previously experienced. So, in January 1939, he and his family made their leisurely way across France, stopping off at some "night spots" in Paris before taking a ship from Marseilles to Alexandria, then a train to Cairo.

The British Air Mission was based at Almaza, alongside the REAF and the Egyptian Flying Club. The aerodrome was still all sand, not yet having hardened runways, and lacked a control tower. Next to the FTS was the REAF's No. 1 Communication Squadron (eventually to include a Royal Flight) and those Flights of No. 4 Squadron which were not posted elsewhere, plus that unit's HQ. There were also workshops for aircraft and engines, while nearby were the offices, hangars and workshops of Misr Air as well as the Egyptian Flying Club. Almaza was already getting crowded.

Weston was now informed by the Chief of the Air Mission that he would be the Technical Advisor to the REAF's Flying Training School. Fred Weston had no authority and could not give orders. He was also told that his position could be delicate, and although he would supervise the technical maintenance of all aeroplanes in the FTS, the Egyptian personnel could choose to accept or ignore his advice. Weston described some of the latter as "anti-British and unhelpful", and he found his lack of executive authority difficult because he had been so used to giving orders. Though the majority regarded Weston as a "colonial mastermind", a few took every opportunity to undermine his status, usually when his advice was based on experience rather than being out of the textbooks. Nevertheless, Fred Weston said that he usually had a good rapport with both his Egyptian and British colleagues, while also getting a local tailor to make him an Egyptian military uniform.

One of Fred Weston's duties at the FTS was to arrange air tests for aeroplanes after they had undergone repair or major overhaul. For this task he was allocated one of the British flying instructors, Bill Brooks, as a test pilot. They usually flew tests together and Bill allowed Fred to handle the controls because he was mad on flying, though official flying instruction was not permitted. Fred Weston therefore progressed to making practice "forced landings" in the desert, out of sight of anyone. One day Bill Brooks suggested that Fred learn to fly officially at the Egyptian Flying Club in another corner of Almaza. Weston did a preliminary flight on an REAF Gipsy Moth then went to the Flying Club for a test flight on one of their Moths, and as a result received his pilot's licence almost immediately. Meanwhile Fred Weston, his first wife Rose and two small sons, Max and Peter, lived in a small hotel in Heliopolis. While in Egypt during the summer of 1973 the author also stayed in a small and by then scruffy hotel in Heliopolis called the *Funduq Tayyarah* (Aeroplane Hotel) where he was told that, in its heyday, this run-down establishment welcomed many airmen as guests, including some from the RAF.

By May 1939 the FTS had 43 Magisters, two Gipsy Moths, 16 or 18 Avro 626, six Audax VIs and 12 Audax Xs. With 79 (or 81) aeroplanes, the FTS was a busy but increasingly crowded place. Fred Weston recalled that one reason why the Miles Magister proved unsuitable in Egypt was that it was almost entirely made of wood, even to the extent of having a thin plywood skin, all of which was vulnerable to wood-boring insects. In some ways the FTS's handful of old, fabric-covered DH 60 Gipsy Moths stood up to the Egyptian climate. However, the arrival of the REAF's Westland Lysanders led to other machines being moved around until 12 "Egyptian" Audax Xs were released for use as advanced trainers. Even so the Advisory Mission's Report of May 1939 emphasised the need for more modern advanced trainers, while the need for spare engines was growing acute.

Almost more serious was the REAF's shortage of fully trained technicians and ground crew, despite the fact that the new Air Mechanics School had passed out its first course of students in January 1939. The latter joined their units in February and proved to be highly competent, especially when led by the REAF's albeit limited number of engineering officers. These men would, for example, solve the Target Towing Flight's biggest problem by making and fitting their own targets and cables. The situation was better at the FTS because here the unit's engineers had been thoroughly trained by the Miles factory which supplied the Magisters. In May 1939 there was also further discussion about the REAF establishing its own aircraft repair depot.

As war approached, plans to replace the REAF's virtually toothless "Egyptian Audaxes" with Blenheim bombers were still under discussion – an idea that would nevertheless be overtaken by events. Early in 1939 negotiations between the Hawker Siddeley Company and the Egyptian government reportedly started again, having failed in 1937. This time Hawker Siddeley offered to construct a factory to assemble Hawker aeroplanes under licence, though the types under consideration remain unknown. Again, the idea was overtaken by war.

Meanwhile Egypt's acquisition of relatively modern military aeroplanes proceeded well, 18 Westland Lysander Mk. Is having been delivered to the REAF's No. 1 Squadron by May 1939, this unit now being designated an army cooperation squadron. They were divided into two Flights, each with five machines, with the remainder held in reserve pending the creation of a third Flight. Two more Lysanders would eventually follow but, as war clouds gathered, No. 1 Squadron suffered from a shortage of spares, especially for its Mercury XII engines, because British units took priority. Also, the Lysander's normal tyre pressures were too hard for conditions in Egypt's Western Desert where more than one aircraft almost nosed over on landing when its wheels dug into the sand of airstrips which lacked hardened runways.

The first Gloster Gladiators to reach the REAF during March and April were immediately sent to No. 2 Squadron whose personnel had been converted into a fighter unit at overcrowded Almaza during February 1939. These Gladiators were Mk. I machines converted to Mk. II standard. From here they went to the RAF's storage unit at Abu Qir in March before eventually reaching the REAF. According to Victor Hubert Tait, Egypt initially had difficulty raising the cash which Britain demanded before handing over its ex-RAF Gloster Gladiator, perhaps accounting for the delay. When these Egyptian Gladiators first appeared in public, they had small oil filters beneath their engine cowlings but these would mostly be replaced by larger sand filters, apparently by the RAF's No. 27 Maintenance Unit.

No. 2 (Fighter) Squadron moved to Dakhailah in March, where it eventually had two Flights of six Gladiators, plus six in reserve. In June the Advisory Mission complained to the British authorities about the slowness of deliveries, whereas the Egyptian Minister of Defence, Sirry Pasha, is said to have asked for deliveries to be delayed because of the REAF's shortage of qualified pilots and technicians. It would seem that the REAF simply could not absorb aircraft fast enough and as a result some Gladiators remained in a storage unit at Abu Qir for use by whoever needed them first – RAF or REAF. According to Tait, further tension was caused by the arrival of Hurricanes for the RAF shortly after the REAF got its first Gladiators, some in the Egyptian Parliament asking why their country had not been offered these clearly more advanced fighters.

Two sleek and modern-looking aeroplanes were delivered to the REAF in May 1939, but these were not fighters to defend Egypt's skies. Instead, they were a pair of twin-engine Percival Q.6 Petrels (serial number Q601 c/n Q43 ex-RAF P5638, and serial number Q602 c/n, Q44 ex-RAF P5638) for Egypt's newly created Royal Flight. They were attached to No. 3 (Communication) Squadron at Almaza and would soon play their part in the war effort. It had originally been intended that King Faruq's Royal Flight would include a De Havilland DH 95 Flamingo but, like so much else, this ambition was overtaken by the outbreak of war.

More Magister trainers were still arriving, a further 25 being delivered to the FTS in February. These presumably included a number of second-hand machines from amongst the ex-RAF serial numbers N3862 to N3866, N3916 to N920, N3875 to N379, N3926 to N3930, N3885 to N3889, N3936 to N3940, N3895 to N3899, N3953 to N3955 and N3912 to N3914, which are known to have eventually been transferred to Egypt. Some replacements for machines written off at the REAF's FTS would also come from RAF stocks, the first of these reportedly being ex-RAF P6418 in June 1939.

By March the frustrating Fairey Gordons formed a Target Towing Flight of four aeroplanes plus two in reserve at Dakhailah, though they remained useless unless towing gear was made. Three were ex-Fleet Air Arm aeroplanes and as yet all they could do was fly over the defences of Alexandria to help Egyptian anti-aircraft gunners and searchlight operators calibrate their guns and lights. There could be no live firing exercises here, nor perhaps in Cairo. In contrast, Egyptian air defence units at Port Said and Suez could call upon RAF Target Towing aeroplanes based at Ismailiya. In May 1939 the Egyptian government had at last placed an order for 18 Blenheim bombers for No. 4 Squadron, to be delivered in October that year, which of course never happened. No. 3 (Communication) Squadron was more fortunate, receiving three second-hand Avro Ansons before war broke out. These appear to have retained their ex-RAF serial numbers: NK581, NK591 and NK271.

Although the REAF was now a separate service from the Egyptian Army, the relationship between the two remained ill-defined. In other respects, outsiders might be forgiven for regarding the REAF as an extension of the RAF, especially as the REAF also introduced the RAF side-cap which, in an Egyptian context, was known as the *Faruqiyah*. It replaced the traditional, and in some respects more practical, *tarbush* or fez. The REAF was also still largely dependent upon RAF stations in Egypt for the more sophisticated aspects of maintenance. On the other hand, the system of promotion in the REAF remained essentially the same as that in the Egyptian Army. Thus, promotions would be made in June and published in July. There were also promotions in January, sometimes as a result of promotions having been delayed for disciplinary reasons but also because the next step in an individual's promotion was made after five and a half years' service.

A new system of ranks was introduced into the REAF early in 1939. Thus, a man was normally promoted from *Tayyar Thani* (Second Pilot, equivalent to the RAF Pilot Officer) to *Tayyar Awal* (First Pilot, equivalent to the RAF Flying Officer) at the end of two years. He would then normally wait three years until promotion to *Qa'id Sirb* (Flight Lieutenant), followed by five and a half years before promotion to *Qa'id Asrab* (Squadron Leader). Each further promotion followed five and a half years to *Qa'id Janah* (Wing Commander), *Qa'id Liwa* (Group Captain), *Qa'id Firqah Jawwiyah* (Air Commodore), *Qa'id Ustul Jawwi* (Air Vice-Marshal) and *Qa'id Asatil Jawwiyah* (Air Marshal). The introduction of these new Air Force ranks apparently resulted in a slight reduction in pay for officers, though other ranks and their pay remained unchanged. There were no serious complaints, though Tait said that some felt disgruntled, perhaps because REAF officers – all of whom were still young – were so proud of being members of this elite service.

Of course, no one in the REAF had yet served long enough to hold any of the senior ranks, apart from King Faruq who assumed the highest rank for himself on suitable ceremonial occasions.

There had been a crisis at the Air HQ in the Egyptian Ministry of Defence early in 1939. Here the most experienced Egyptian Army staff officer was removed for what the British Advisory Mission described as political reasons and was replaced by a pleasant but very inexperienced man, again from the Army because no REAF officer was senior enough. The British declared him to be ineffective, though this seems to have overstated the matter. Nevertheless, his position was unenviable because the Egyptian Air Force wanted one of their own to be Chief of the Air Staff.

On a more positive note, the REAF took part in large scale joint exercises by Egyptian and British forces lasting from late February to April 1939. Designed to improve the defence of the Suez Canal, these involved the British Canal Brigade based at Fayid Camp and the Egyptian Army's Nos. 4 and 5 Battalions under *Liwa* (Brigadier) Sayf Bey, plus Egyptian Army artillery and the REAF. During one part of these exercises a Flight from No. 1 (Army Cooperation) Squadron was based at Suez to carry out reconnaissance of the Canal area, soon demonstrating that No. 1 Squadron was the most efficient unit in the REAF.

An exercise in the Western Desert was especially satisfying for the Egyptians as it involved the Egyptian Army's Armoured 1st Car Regiment (or 1st Armoured Car Regiment) at Bahariya oasis, two squadrons of the Frontier Force, a section of anti-tank guns and a Flight of REAF Lysanders from No. 1 Squadron. It was partly intended to test the "Combined Plan for the Defence of Egypt", in which the Egyptian Army's role was to patrol the western frontier, provide a small mobile force to patrol the desert south-west of Cairo and provide Frontier Force units to protect the Alexandria to Marsa Matruh coastal railway. The most successful aspect was a night drive by the Frontier Force from Bahariya to Siwa Oases, along the southern fringe of the daunting Qattara Depression, without lights. Meanwhile the Flight of No. 1 Squadron Lysanders practised reconnaissance and bombing in the area of Bahariya while also working in support of Frontier Force armoured cars.

Another exercise focussed upon the air defence of Alexandria. Named the Alexandria Manning Exercise, it demanded close cooperation between Egyptian and British units, the Egyptian anti-aircraft proving notably effective during mock attacks upon the city, harbour and naval facilities. However, the Egyptian searchlight crews were found to need more training. Despite the fact that No. 2 (Fighter) Squadron was still forming at Dakhailah, nine of its Gladiators took part, as did No. 4 Squadron's old Audaxes which played the part of "enemy". The Coastal Artillery Batteries now established at Marsa Matruh and Alexandria were similarly involved.

The last of this series of urgent exercises again involved the Egyptian Frontier Force and Cavalry whose Nos. 1, 2, 3 and 4 Bren Gun Carrier squadrons manned a temporary Air Observation Line across the Western Desert. The Frontier Camel Corps had been re-equipped with lorries in order to man the Observation line while four companies of the Egyptian Army's Arab Legion carried out this and other duties at Marsa Matruh, Sidi Barrani, Siwa Oasis and Bahariya Oasis. Labour companies from the Egyptian Army's Engineers similarly worked at Marsa Matruh as part of the exercise.

Even before these exercises came to an end, the popular but somewhat bumbling *Liwa* (Brigadier) Ali Islam Bey was replaced in May 1939 by *Liwa* Hasan Abd al-Wahab Pasha, an artillery officer who had commanded Frontier Force units during the exercise in the Western Desert. A stern disciplinarian with staff experience, Abd al-Wahab Pasha was regarded by the British as a great improvement. However, he was still an Army rather than an Air Force man and many in the REAF regarded him as a "palace placement" or even as "King Faruq's spy". Perhaps in an attempt to counter such views, Brigadier Hasan Abd al-Wahab was allowed to adopt the comparable Air Force rank of *Qa'id Ustul Jawwi* or Air Vice-Marshal. Much later *Qa'id Asrab* (Squadron Leader) Abd al-Muna'im al-Miqaati told the author that Ali Islam had been popular in the REAF because he spoke up for Air Force officers. In fact, they suspected that Ali Islam had been 'promoted out of the REAF' by King Faruq for speaking up too much.

Liwa Abd al-Wahab would remain in charge until late in 1940 when he reportedly resigned because he could not get on with the Egyptian Army's new Chief of Staff. Despite his age, he also used this period to try to learn to fly and get REAF wings but, after training on Miles Magisters at the FTS in Almaza, Abd al-Wahab was unable to reach the standard required of an REAF pilot. In practice *Qa'id Asrab* (Squadron Leader) al-Miqaati, one of Egypt's first three military pilots, already ran the REAF despite his relatively lowly rank meaning that he nominally remained only the Commander of Almaza aerodrome. Here he was the immediate superior of Abd al-Hamid al-Dighaydi and Nu'man Nada who were in command of the FTS. In a note written in June 1939, Group Captain Victor Hubert Tait stated that 'the way in which the Egyptian officers have managed to cope with their heavy responsibilities in spite of their lack of administrative experience was altogether admirable'.

Sadly, politics again interfered in the affairs of the Egyptian Army and Air Force in 1939. The British Advisory Mission had already been disturbed by an apparent change in the attitude of Husain Sirry Pasha, the Egyptian Minister of War who, having previously been very helpful, suddenly became less so. They suspected that it reflected pressure from the palace. Then, with the fall of Muhammad Mahmud Pasha's government, Sirry Pasha was removed by the new Egyptian Prime Minister, Ali Mahir, who came into office on 18 August and would remain until 28 June 1940 when he would be ousted on British insistence.

On the other hand, the Advisory Mission regarded the new Minister of War brought in by Ali Mahir as an improvement. Ali Mahir also confirmed General Aziz al-Masri as the Army's new Chief of Staff, having previously been the nominal head of the Egyptian Army following the Anglo-Egyptian Treaty of 1936. The British then realised (or claimed to have suddenly discovered) that al-Masri had a German secretary, so there was consternation at the British Embassy. Furthermore, General Aziz al-Masri divorced his American wife in 1939. In reality the fuss was something of a farce, as the female secretary in question proved to be Jewish and, of course, profoundly anti-Nazi. Though General al-Masri remained little more than a figurehead, even the British acknowledged that his tour of the Western Desert oases shook up local officials and resulted in a useful report. British paranoia was further fuelled by a rapprochement between King Faruq and his cousin, Prince Abbas Halim whom the British Embassy still insisted on describing as somehow pro-German.

The continuing role of the REAF in combatting the scourge of drug smuggling seems almost insignificant in comparison to the political intrigues which rocked Cairo in the months leading up to the Second World War. Nevertheless, this important work

Egyptian Prime Minister Muhammad Mahmud Pasha [centre], accompanied by the Minister of Finance Dr Ahmad Mahir [left], the Commander of the Egyptian Army [right] and other senior figures, watching an air display at the REAF Flying School at Almaza in 1938. (Author's collection)

went on because the French mandate authorities were finding it almost impossible to stamp out the growing of opium poppy in mountainous districts of Syria and Lebanon. In fact, the final phase of the anti-drug smuggling campaign started in 1939 when herds of camels, which were being imported into Egypt for meat, had containers of drugs forced down their throats. The animals were then brought across the frontier quite openly. Up to 27 containers were found forced in one animal's stomach and this method of smuggling was only defeated when X-ray equipment came into use.

The last of the British Advisory Mission's reports on the REAF before the start of the Second World War was submitted at the end of April 1939. It stated that the Egyptian Air Force now had 18 Avro 626s, 43 Magisters, 24 "Egyptian Audaxes" (6 Audax IVs and 18 Audax Xs), 18 Lysanders, 18 Gladiators, six Gordons, two Gipsy Moths, one Avro 652A, one Wessex and one Avro 641 Commodore. Of this total of 132 aeroplanes, only 36 were front-line machines.

A supplementary report in May noted that a second squadron of REAF Gladiators was expected in July and that No. 1 (Army Cooperation) Squadron now had two Flights of Lysanders, plus four in reserve. Unfortunately, this unit was still short of engine spares and continued to face difficulty with high-pressure tyres when landing in sandy deserts. No. 2 (Fighter) Squadron was currently forming at Dakhailah and consisted of two Flights, each with six aeroplanes plus three in reserve. However, no spares had arrived. No. 3 (Communication) Squadron remained unchanged but expected two Percival Q.6s and a De Havilland Flamingo for the Royal Flight (the latter never arriving). No. 4 (Bomber) now only had one Flight which consisted of "Egyptian Audax" Xs which were suffering from a serious shortage of spares, although 18 Bristol Blenheims had been ordered (which also never arrived).

The Signals station (presumably the highly secret "remove radar station") on the road to Suez was now operational and a temporary Air Observation System had been tried out in air defences exercises with the British, involving mock attacks upon Alexandria. Airfields were already available at Daba'a, Bir Hukir, Hammam, al-Qasaba, Burj al-Arab and Maryut. Of these locations, Daba'a had been selected because it had a secure water supply, lay on the coast in favourable terrain and could make it easier for British forces in the Egyptian heartlands to reach Marsa Matruh quickly in case of a threat from the Italians in Libya. Work had also started on providing Marsa Matruh airfield with permanent structures while other work was being carried out on airfields at Fuka, Khataba, Suez, Bahariya and al-Tur.

Almost forgotten in the deep south of the country other Egyptian and British officers checked the defences of the Aswan area, still fearing an Italian raid in the event of war. What these exercises had shown, sometimes in a brutal manner, was that the Egyptian Army was not in as good shape as the Egyptian Air Force. In January 1939 a report by the British Advisory Mission had highlighted poor training, inadequate senior officers and poor standards of maintenance. This was when the idea of establishing an Arab (Bedouin) Corps of four companies (600 men) to defend the Western Desert oases was proposed. Officers of Arab origin were, in fact, already being selected for a new Arab Battalion and it was hoped that the sons of local Bedouin shaykhs would volunteer for the role. However, it remained clear that there were not enough Egyptian anti-aircraft units – in terms of officers or of men – to be sent to the Sudan, as they were all needed within Egypt.

The Army's 1st Light Tank Regiment (still nominally regarded as cavalry) was at half strength but had completed its training. In contrast, the 1st Light Car (mechanised but unarmoured) Regiment was almost untrained. The Egyptian Army also retained one cavalry unit mounted on horses, largely for ceremonial purposes. As far as infantry were concerned, the 1st Brigade was based in Cairo, the 2nd Brigade at al-Miaks outside Alexandria, and the 3rd Brigade at Manqabad where it also had the keenest most efficient officers. A substantial proportion of Egyptian Army Engineers was working on the construction of fixed defences around Marsa Matruh. With five squadrons, the Frontier Force was almost up to full strength, training well and would hopefully soon be joined by the Arab Battalion.

In February a new Five Year Plan was introduced by the Egyptian government and although this did not offer much to the infantry (currently nine rifle battalions in Egypt and one in the Sudan) and machine gun units, the anti-aircraft, coastal defence batteries and to a lesser extent the mechanised "cavalry" would fare better. Such priorities reflected what the British wanted, but in March and April the British were taken aback when the Egyptian Army ordered 1,000 Bren light machine guns directly from what had been Czechoslovakia. In fact, the rump of Czechoslovakia (its frontier provinces having been annexed by Nazi Germany and

Poland between September and November 1938) was invaded by Germany on 15 March 1939, resulting in the creation of the Nazi Protectorate of Bohemia and Moravia, and a puppet Slovak State. British official documents stated their concern that these weapons were not compatible with the version of the Bren used by British forces, reportedly fearing the guns might get mixed up and thus lead to confusion. Also, in April the Egyptian Army was invited by the British Army to set up its own anti-aircraft artillery batteries in Suez and Port Said, which it soon did.

During this tense period Report Number 9 on the Egyptian Army by the British Advisory Mission described the situation as 'both heartening and disappointing' at the same time. Significant improvements had been made rapidly and Egyptian infantry were capable of cooperating efficiently with British units. The anti-aircraft artillery units proved better than expected during the Alexandria air defence exercise and were well up to British standards, though the searchlights were not. The discipline of ordinary soldiers was good and was indeed better than that of their officers or technicians, but there was still a desperate shortage of equipment for the Artillery, Engineers and Signals. Much to the relief of the British the proposed purchase of Czech Bren guns had fallen through.

However, a traditional preoccupation with appearances over substance sometimes still had a detrimental effect. For example, during the rehearsal for a recent military parade to commemorate the marriage of the King's sister, Princess Fawzia Fu'ad to Muhammad Reza Pahlevi, the Crown Prince of Iran, King Faruq told a dozen of the fattest senior officers in the Army that they could not take part in the march-past. This included *Liwa* (Brigadier) Sayf Bey who, overweight as he might have been, was regarded by the British as the Egyptians' best infantry officer. Of those senior officers who were permitted to take part in the parade, the best was, in the eyes of the British, Prince Isma'il Da'ud. He had only recently returned to service in the cavalry as a *Bimbashi* (Major) after 20 years' absence and was again far from slim. However, Isma'il Da'ud was a member of the royal family and would soon be promoted in rank. Rather to the surprise of the British, Prince Isma'il Da'ud would also prove to be a highly competent commanding officer. This parade included a section illustrating the "History of the Egyptian Army" which included men wearing early 19th century uniforms from the time of Muhammad Ali, the founder of the Egyptian royal dynasty. They were probably the reconstructed uniforms still displayed in the Cairo Citadel's Military Museum.

Far to the south, in the Anglo-Egyptian Sudan, the Egyptian 7th Infantry Battalion at Port Sudan and Khartoum was described by the Advisory Mission as having very low efficiency, though the British General Officer commanding Sudan was going to arrange further training for them. Four Egyptian officers and 80 soldiers were also to be sent to Port Sudan to man the Coastal Artillery. According to Anwar Sadat, his friend and colleague Gamal Abd al-Nasser volunteered to go with the 3rd Battalion to Khartoum a short while later. There he would meet his future colleague and revolutionary, Hakim Amr. This infantry battalion had previously been stationed at Manqabad in Upper Egypt, where it was described by the British Advisory Mission as the best in the Egyptian Army.

On 21 March 1939 a report on the recreation of an Egyptian Navy proposed a total of nine ships, to consist of one large escort vessel, four minesweepers and four Motor Torpedo Boats (MTBs). Almost exactly a month earlier, Captain G.T. Philip of the British Royal Navy had arrived in Egypt to serve as Naval Advisor to the Egyptian Ministry of Defence. He soon put forward a modified version of the earlier plan, suggesting a smaller escort vessel, six large submarine chasers instead of the four MTBs and just two minesweepers. Two of the submarine chasers and the minesweepers would be built in the Alexandria dockyard under technical direction from Thorneycroft, the famous English naval shipbuilder. There was also considerable emphasis on the establishment of an anti-submarine "Indicator Loop" or submerged cable to be laid about 13km out to sea, as part of Egypt's coastal defences. Egypt would have naval bases at Alexandria, Suez and Port Said. For administrative purposes it was suggested that Egypt's existing Marine Department of Coastguards, Ports and Lights Authority and Royal Yachts be united under the Ministry of Defence.

The Egyptian Coastguard already had two small transport ships, the *Al-Amira Fawzia* and the *Sollum*, plus the patrol boat *Al-Amir Faruq*, and it was suggested that the latter's armament of a single 6-pdr gun be increased. Then there was the slow but still useful patrol vessel *Mahabiss* which was currently used in fisheries research. Two other existing vessels, the *Niphtis* and *Al-Amir Abd al-Muna'im*, were, however, too old to be useful. A small number of naval personnel had been recruited and their training officially started on 17 March 1939 though four Royal Navy minesweepers had been teaching a handful of Egyptians in Alexandria since February 1938. Two more vessels were now arriving and Admiral Pound, commander of the Royal Navy's Mediterranean Fleet, offered these to Egypt as well. The Egyptian government was eager to accept any help in terms of ships or training that the British cared to offer but in practice most training was the responsibility of *Qa'im Maqam* Browning of the Egyptian Coastguards Service, the only official in Egyptian service who had real knowledge of the sea. He and Captain Philip would, it was anticipated, 'start the Egyptian Navy'.

For reasons which remain unclear, the British authorities were in a flap in 1939 about a proposed overseas flight by Hassan Anis Pasha, the head of civil aviation in Egypt. Egypt was, of course, on tenterhooks through that summer, fearing war but hoping it could be avoided, though in reality neither Egypt nor its small and still ill-equipped armed forces could do much to influence events. Adolf Hitler was clearly bent on conflict and it was a matter of when rather than if war began. In August 1939, in the final weeks of peace, the air defence of Egypt was in the hands of three British fighter squadrons, supported by one recently operational Egyptian fighter squadron with another in the process of formation. On 2 August, plans to expand Egypt's Army, which had been held up by a shortage of money, seem to have restarted – at least in part.

August also saw the last report (Number 10) by the British Advisory Mission before the outbreak of war. Patronising in tone, it nevertheless described itself as 'an encouraging report', the Egyptian Army being adequate if not brilliant. 'The anti-aircraft regiment is really good', while all other formations had increased in efficiency with the notable exception of the Light Tank Regiment which had what the British advisors described as a thoroughly bad CO who only retained his position through political influence. The results achieved by the 1st Anti-Aircraft Regiment were as good as those of a regular British unit while a Coastal Defence Unit was 'up to Malta standards', and had recently been sent to Port Sudan. Unfortunately, a recent Cairo Manning Exercise to assess the air defences of Cairo had not been as successful as the previous Manning Exercise at Alexandria.

A pair of Bloch MB 200 heavy bombers of the French Armée de l'Air's GB I/6 escadrille VR 552 based at al-Blidah [Blida] in flight near the Algerian coast shortly before the Second World War. A slightly newer design that the Farman F.221, it was still outdated and it was being phased out of front-line duties when the war broke out. (Remi Baudru via Jarrige)

Then there had been a near riot at the garrison town of Manqabad in Upper Egypt when reservists were told they would be kept in service for a further two weeks. This resulted in what the British described as the imposition of savage discipline by the respected if not much-liked *Liwa* Sayf Bey, the Egyptian commanding officer at Manqabad. Since then, there had been no further trouble, though something similar had occurred earlier in Alexandria. Sayf Bey was also hated by Wafdist politicians in the Egyptian Parliament.

A great effort had been made to increase the size and efficiency of the Egyptian tank units but this would take time. Meanwhile the elite Frontier Force now had five Light Car and Mounted squadrons, which was a significant expansion achieved despite a shortage of British advisors. Some Egyptians might suggest that it had been achieved because of a lack of British interference. Another potential difficulty, which came to nothing because of the outbreak of war, stemmed from a proposal to send Egyptian officers to Turkey for training when the British were unwilling or unable to accept more courses in Britain. Nevertheless, the Advisory Mission anticipated such a policy causing confusion. August 1939 similarly saw the Egyptian Army's 1st Anti-Aircraft Regiment reaching a fully operational state.

While all this was going on, No. 208 (Army Cooperation) Squadron RAF moved to Marsa Matruh on 7 August as part of a scheme developed by HQ Egypt Group whereby all squadrons moved to their war stations. However, British Army requirements and the location of the Headquarters of the Armoured Division, meant that No. 208 Squadron had to move back to al-Qasaba on 1 September, the very day that Germany invaded Poland. This confusion inevitably caused No. 208 Squadron's training to be seriously interrupted but fortunately for the British, the French and indeed the Egyptians – though not of course for Poland – the outbreak of the Second World War would be followed by what came to be known as "The Phoney War".

BIBLIOGRAPHY

PUBLISHED BOOKS

Anon., *Tarikh al-Quwat al-Iraqiyat al-Musalat. Al-Jazz' al-Saba 'Ashar. Ta'sus al-Quwwat al-Jawiyah wa Tatwirha*, in Arabic [History of the Iraqi Armed Forces, Part 10. Establishment and Development of the Air Force] (Baghdad: 1988)

Berque, J., *Egypt, Imperialism and Revolution* (London: Faber & Faber, 1972)

Bou-Nacklie, N.A., *Les Troupes Speciales du Levant: Origins, Recruitment and the History of the Syrian-Lebanese Para-military Forces under the French Mandate, 1919-1947* (Ann Arbor, 1989)

Butt, G., *History in the Arab Skies. Aviation's Impact on the Middle East* (Nicosia: Rimal Publications, 2011)

Cooper, T., & M. Sipos, *Wings of Iraq Volume 1: The Iraqi Air Force, 1931-2003* (Warwick: Helion & Company, 2020)

Gavin, R.J., *Aden under British Rule, 1839-1967* (London: 1975)

Goldschmidt, A., Johnson, A.J. & Salmoni, B.A., *Re-Envisioning Egypt 1919-1952* (Cairo: American University in Cairo Press, 2000)

Halim, N. Abbas, *Diaries of an Egyptian Princess* (Cairo: Zeitouna, 2009)

Jabr, Jabr 'Ali [Gabr, Gabr Ali], *Al-Quwwat al-Jawiyah bayn al-Siyasat al-Misriyah wa'l-Isra'iliyah, al-Jazz' al-'Awal 1922-1952*, in Arabic [The Air Force between Egyptian and Israeli Policies, 1922-1952] (Cairo 1993), volume 1

Kelly, S., *The Lost Oases: The desert war and the hunt for Zerzura. The True Story Behind "The English Patient"* (Boulder Colorado: Westview Press, 2003)

Labib, Ali Muhammad, *Al-Quwat al-Thal☐thah*, in Arabic [The Third Force] (Cairo: 1977)

Marr, D.S.B., *A History of 208 Squadron* (Southend 1966)

McGregor, A., *A Military History of Modern Egypt, From the Ottoman Conquest to the Ramadan War* (Westport: Praeger, 2006)

Nordeen, L., & Nicolle, D., *Phoenix over the Nile. A History of Egyptian Air Power 1932-1994* (Washington: Washington Institution Press, 1996)

Nowarra, H.J., *Junkers Ju 52* (Sparkford: Haynes, 1987)

Raafat, S.W., *Privileged for Three Centuries. The House of Chamsi Pasha* (Cairo: Samir W. Raafat, 2011)

Stuart-Paul, R., *The Royal Saudi Air Force: a legacy of His Majesty King Abdul Aziz Ibn Saud* (London: Stacey International, 2001)

Wenner, W.M., *Modern Yemen 1918-1966* (Baltimore: Johns Hopkins University Press, 1967)

Weston, F., *A Trenchard Brat: A Life in Aviation* (Studley: Brewin Books 1999)

JOURNAL ARTICLES

Andersson, L., 'Wings Over the Desert: Aviation on the Arabian Peninsula. Part One Saudi Arabia', *Air Enthusiast*, 112 (July-August 2004), pp.39–43

Anon., 'How did you join the Air Force?' [recollections of Air General Mahmud Sidqi], *Al-Quwat al-Musalah* (November 1959), pp.22–23

Garello, G., 'Ali Italiane sull'Iraq (1937-1941)', *Aero Fan*, 24 (March 2006), pp.2–23

ONLINE SOURCES

Andersson, L., 'Wings Over the Desert, Aviation on the Arabian Peninsula, Part 2', <http://www.artiklar.z-bok.se/Arabia-1.html>

UNPUBLISHED SOURCES

Abu Zaid, Muhammad Abd al-Hamid, Pilot's logbook from 22/2/1939 to 31/8/1944 (transcribed by the author)

Anon., 'Biographical Note on the late Squadron Commander Mohamed Abdul Hamid Abu Zeid' (EAF Historical Department, n.d., supplied to the author 2001)

Cooper, T., Correspondence concerning the Royal Iraqi Air Force from the 1930s to the 1950s (1997 to 2015)

Gabr, Gabr Ali, Air Brig. Dr, 'The Arab Israeli Conflict: The Roots and the Wars 1897-1979: An Egyptian Perspective' (unpublished book text, 2007)

Gazerine, I.H., Group Capt., Interview by the author (Cairo, 1973); correspondence following (1973–1974)

Kafafi, M., Interview by the author concerning her father, Adli Kafafi (Cairo, 1999)

Miqaati, Muhammad Abd al-Muna'im al-, Air Commodore, Interview by the author (Cairo, 1973); correspondence (1974)

Miqaati, W. al-, Correspondence concerning his father, Abd al-Muna'im al-Miqaati (2013 to 2019)

Sabry, Fu'ad, Interview by the author concerning members of his family in the REAF and EAF (Cairo, 2011)

Sherif Abu Zeid, Interview by the author concerning his uncle, Abd al-Hamid Abu Zaid (Alexandria, 2000)

Sidqi, Mahmud, Air Marshal, Anonymous interview in the late 1950s (text supplied by Nour Bardai)

Stafrace, C., 'The Iraqi Air Force' (pre-publication draft supplied by the author, 2008)

Tait, V.H., Air Marshal, Interview by the author (London, 1974)

Tawfiq, Hassan, Pilot's logbook 1935-50 (selected pages supplied to the author)

Tewfik, M., Interview by the author concerning her father Hassan Tawfiq (London, 1997 and 1999)

Zaki, Brig. Dr Abd al-Rahman, Interview by the author (April 1973)

Zaki, Brig. Dr Abd al-Rahman, 'An Adventurous Event & Career' [outline of the life of Gen. Aziz al-Masri] (unpublished text, supplied to the author, 1977)

THE NATIONAL ARCHIVES (LONDON)

AIR 2/2768: First half yearly report on the REAF, 1937

AIR 2/3777: RAF NCOs seconded to RIrqAF, 1938–1941, conditions of service, etc., January 1941

AIR 2/3811: RAF personnel seconded to British Military Mission Iraq, 1938–1939

FO 371/20912: Report on the EAAF and Egyptian Army by British Ambassador in Cairo for Foreign Office, with supplementary note by Major General Marshall-Cornwall, 1937

FO 371/21852/433: On RAF personnel for RIraqAF, 1938

FO 371/21939: Report on REAF by British Advisory Mission for Foreign Office London, 24 May 1938

FO 371/21944: Report on REAF by British Advisory Mission for Foreign Office London, 24 May 1938

FO 371/23326: Combined Plan for the Defence of Egypt, report for the Foreign Office London, August 1939

FO 371/23331: Report on REAF by British Advisory Mission for Foreign Office London, 14 May 1939

FO 371/23334: Report No. 10 on Egyptian Army by British Advisory Mission for Foreign Office London, August 1939

FO 371/23337: Report on REAF by British Advisory Mission for Foreign Office London, 30 November 1939

FO 371/24610: Foreign Office comments added to Report No. 11 on the Egyptian Army by the British Advisory Mission for Foreign Office London, December 1939

ABOUT THE AUTHOR

Dr David C Nicolle is a British historian specialising in the military history of the Middle Ages, with special interest in the Middle East and Arab countries. After working for BBC Arabic Service, he obtained his MA at SOAS, University of London, and a PhD at the University of Edinburgh. He then lectured at Yarmouk University in Jordan. Dr Nicolle has published over 100 books about warfare, mostly as sole author and co-authored the Arab MiGs series on the history of the Arab air forces at war with Israel.